Data-Driven Innovation

Today, innovation does not just occur in large and incumbent R&D organizations. Instead, it often emerges from the start-up community. In the new innovation economy, the key is to quickly find pieces of innovation, some of which may already be developed. Therefore, there is the need for more advanced means of searching and identifying innovation wherever it may occur.

We point to the importance of data-driven innovation based on digital platforms, as their footprints are growing rapidly and in sync with the shift from analogue to digital innovation workflows. This book offers companies insights on paths to business success and tools that will help them find the right route through the various options when it comes to the digital platforms where innovations may be discovered and from which value may be appropriated.

The world hungers for growth, and one of the most important vehicles for growth is innovation. In light of the new digital platforms from which data-driven innovation can be extracted, major parts of analogue workflows will be substituted with digital workflows.

Data-driven innovation and digital innovation workflows are here to stay. Are you?

Michael Moesgaard Andersen is Adjunct Professor of Strategy and Innovation at Copenhagen Business School in Denmark. He runs his own venture-capital company, which invests in entrepreneurial tech companies dealing with AI and software-based solutions. He has previously served as the owner and CEO of a globally focused consulting business, a partner at Deloitte, and a civil servant in the Danish Ministry of Finance.

Torben Pedersen is Professor of International Business at Bocconi University in Milan, Italy. His research focuses on the interface between strategy and international management. He has published more than 100 articles and books as well as more than 25 teaching cases. In addition, he has directed a number of research projects and is one of the leaders of the Manufacturing Academy of Denmark. He is Fellow of the Academy of International Business, the European International Business Academy, and the Strategic Management Society.

"Will big data, AI, and network externalities result in the growing dominance of a handful of technology monopolists led by Alphabet, Apple, Amazon, Alibaba and Tencent? Andersen and Pedersen offer an alternative vision: start-ups and smaller firms revitalizing economic performance through embracing new digital workflows, cloud computing, software-as-a-service, and multi-user platforms to deliver data-driven innovation."

— **Robert M. Grant**, Emeritus professor of management, Bocconi University

"Invincible companies need to continuously reinvent themselves in a systematic and repeatable manner. This book highlights how data-driven innovation can help fuel your innovation pipeline with start-ups in the field of new digital platforms."

— **Alexander Osterwalder**, Dr., inventor of the Business Model Canvas, and founder of Strategyzer

"The world desperately needs more innovation and entrepreneurship, both to solve its pressing problems and to achieve broadly shared economic growth and prosperity. The insights contained in *Data-Driven Innovation* can help entrepreneurs and those who want to support and partner with them to identify the start-ups with the highest potential to blitzscale their way to global impact. Readers will learn how to innovate or acquire innovation more quickly, and with a much higher hit-rate."

— **Chris Yeh** is the cofounder of Global Scaling Academy and founder of Blitzscaling Ventures

"*Data-Driven Innovation* is a thorough and practical guide for executives everywhere, as they seek to get to grips with the opportunities afforded by the digital revolution. Michael Andersen and Torben Pedersen bring a unique combination of practice experience and academic insight to tackle this hugely important issue."

— **Julian Birkinshaw**, Professor of Strategy & Entrepreneurship, London Business School

"Over the last 10 years, I have gone from being somewhat sceptical of some elements of the data-driven innovation concept as described in this book to being extremely enthusiastic. It is highly recommended."

— **Lars Tvede**, venture capitalist, serial entrepreneur, and author of 14 books

"The data-driven economy is here to stay – and it has a profound impact on how we design innovation strategies and how we are able to engage in innovation activities. Here's an indispensable and innovative guide not only for R&D managers but everyone who is engaged in innovation!"

— **Marc Gruber**, Professor for Entrepreneurship & Technology Commercialization, VP Innovation at the Swiss Federal Institute of Technology, Lausanne

Data-Driven Innovation

Why the Data-Driven Model Will Be Key to Future Success

Michael Moesgaard Andersen
and Torben Pedersen

Routledge
Taylor & Francis Group

LONDON AND NEW YORK

First published 2021
by Routledge
2 Park Square, Milton Park, Abingdon, Oxon OX14 4RN

and by Routledge
52 Vanderbilt Avenue, New York, NY 10017

Routledge is an imprint of the Taylor & Francis Group, an informa business

British Library Cataloguing-in-Publication Data
A catalogue record for this book is available from the British Library

Library of Congress Cataloging-in-Publication Data
Names: Andersen, Michael Moesgaard, author. | Pedersen, Torben, author.
Title: Data-driven innovation : why the data-driven model will be key to future success / Michael Moesgaard Andersen and Torben Pedersen.
Description: Abingdon, Oxon ; New York, NY : Routledge, 2021. | Includes bibliographical references and index.
Identifiers: LCCN 2020043431 (print) | LCCN 2020043432 (ebook) | ISBN 9780367485771 (hbk) | ISBN 9781003041702 (ebk)
Subjects: LCSH: Business—Technological innovations. | Success in business. | Big data—Economic aspects.
Classification: LCC HD45 .A61724 2021 (print) | LCC HD45 (ebook) | DDC 658.4/063—dc23
LC record available at https://lccn.loc.gov/2020043431
LC ebook record available at https://lccn.loc.gov/2020043432

ISBN: 978-0-367-48577-1 (hbk)
ISBN: 978-1-003-04170-2 (ebk)

Typeset in ScalaSans
by Apex CoVantage, LLC

Contents

Figures and tables

Figures

Tables

Preface

Today, data-driven innovation is the exception. Tomorrow, data-driven innovation will be the norm. This change is the reason for this book.

Why has this radical change not already occurred?

For many years, innovation took place behind closed doors. The underlying justification for such secretive innovation was implicitly recognized by Machiavelli in his masterpiece, *The Prince*, where he wrote:

> It ought to be remembered that there is nothing more difficult to take in hand, more perilous to conduct, or more uncertain in its success, than to take the lead in the introduction of a new order of things. Because the innovator has for enemies all those who have done well under the old conditions and lukewarm defenders in those who may do well under the new. This coolness arises partly from fear of the opponents, who have the laws on their side, and partly from the incredulity of men.[1]

Machiavelli highlights the classic resistance to a new order of things. In this context, it is only natural to expect resistance to fast digital platforms that provide access to data-driven innovation when we are accustomed to slow, analogue-based workflows.

Another reason why data-driven innovation has yet to have a proper breakthrough may be related to the common and widely recognized risk of failure when switching from well-tested techniques and systems to new procedures. Why run the risk if the innovations may not have the intended effect? Even if they do, why would you subject yourself to cannibalism? In other words, why risk cutting off the branch on which you sit so comfortably?

Moreover, many people fail to recognize the fact that access to innovation based on digital platforms can now be purchased. Most people believe that innovation is something that an organization has to create itself, and if an organization wishes to acquire innovation, it must acquire another company.

On this basis, we have taken the liberty to write this book with the aim of turning attention to new ways of identifying, attracting, and applying innovation. We believe that a new wave of innovation is building which will eventually replace closed and open innovation. This new wave of data-driven innovation will allow you to search globally for innovative ideas and not just among your close collaborators. New and innovative ideas might pop up in places you never dreamed about, and you can expand your search for these novel ideas to the whole world.

The future of innovation will involve data-driven innovation based on new digital platforms. As we dug through the world of data-driven innovation, we found a number of interesting and varied digital platforms. Therefore, this book sets out to explore and analyze these platforms so those who are ready can gain firsthand insights into the many new opportunities to purchase access to innovation. In some instances, it may be enough to subscribe to just one platform, while in others it may be necessary to acquire innovation from several platforms. There is no such thing as "one size fits all" at this moment. However, at some stage we may see a universal platform that connects all of the constituents in the innovation economy—just like LinkedIn became the sole platform for professional networking, and Airbnb became the go-to platform for home sharing.

Throughout our exploration of these platforms, we were also interested in ways of exploiting innovation. In other words, how do you appropriate value from innovation? This interest made us aware of the assumed inherent dichotomy between innovation and exnovation— you gain access to innovation, after which you standardize it and scale it up (exnovation). Notably, we found that the timeline does not necessarily have two consecutive phases. On the contrary, we found an ambidextrous relationship between innovation and exnovation. This intimate relationship is important because innovation is largely irrelevant if it is not transformed into growth and value appropriation. Nevertheless, the main focus in this book is on data-driven innovation.

We are indebted to everyone who has assisted us in the journey toward this final product. Heartfelt thanks go to our many inspiring colleagues at the Strategic Management Society, Bocconi University, and Copenhagen Business School. Particularly helpful and beneficial input came from numerous colleagues from the academic world and the many executives with whom we have had many stimulating dialogues. We would also like to thank our many reviewers, including those who provided peer reviews of our book proposal and final reviewers like Christian Lawaetz Halvorsen, who helped us with review of the AI technology aspects of the book. Our gratitude also goes to Martin Fenge, who designed the cover and provided the graphics.

Special thanks go to the publishers at Routledge, notably senior editor Rebecca Marsh who assisted with swift response and constructive guidance regarding all our tricky questions.

Moreover, we also owe thanks to Jette Sørensen for diligently orchestrating the administrative and secretarial tasks related to the publication of this book. Finally, a special thanks goes to our families who fully believed in our passion for writing this book, notably Asta Moesgaard Andersen who even came through with invaluable support as to how we can reinterpret Socrates when it comes to understanding human cognition of innovation.

Michael Moesgaard Andersen and Torben Pedersen

Copenhagen Business School and Bocconi University

July 2, 2020

Note

1 See Machiavelli, N. (2010). *The Prince*, edited by Michael Ashley. Campbell, CA: FastPencil Inc, p. 21.

chapter 1

Searching for accelerated growth in a data-driven world

Since the beginning of the 2008–2009 financial crisis, major parts of the Western world have suffered from lower growth. Moreover, somewhat satisfactory growth on the global scene has been skewed by the hypergrowth exhibited by the BRIC (Brazil, Russia, India, and China) countries, which was well above the global average, leaving a large number of Western countries far behind with growth figures significantly below the global average. This lack of macroeconomic growth escalated further with the emergence of the COVID-19 crisis in 2020. Global GDP dropped dramatically, and the repercussions of the crisis extended well into the ensuing quarters. In general, stimulus packages involving various combinations of quantitative easing and negative interest rates from the US Federal Reserve have not been sufficient remedies for the challenges in recent years.

On the microeconomic level, many companies, including large global corporations, are suffering from a lack of growth. Consider for instance, General Electric, Exxon Mobile, Pfizer, CitiGroup, and Walmart, which were the world's five most valuable companies in 2000. Today, they have been surpassed by high-growth companies such as Amazon, Google (Alphabet), Microsoft, Apple, and Facebook from the United States, and Tencent and Alibaba from China. Most of these companies (with the exception of Microsoft) did not even exist in 2000. This leads us to a key question: What characterized the old leaders, and what characterizes the new?

Invariably, the business models of the old leaders were tied to physical products and/or physical assets. The new leaders are IT companies in which growth is driven by data, especially big data. Big data allows for innovation as new services are enabled solely on the basis of data. Consider for example, Google and Facebook, who are masters in converting "thin" data generated on their platforms without cost into "thick" data of high value for advertising, analytics related to consumer behavior, and other types of analytics. While thin data are collected for free, thick data are sold for millions.

In other words, the new leaders are utilizing data innovatively. Another approach is to use data to search for and access innovation wherever you might find it. Approaching innovation in a data-driven fashion is the topic of this book, which focuses on new digital platforms that can deliver such data and help transform innovation processes into digital workflows.

On the macroeconomic level, the term "data-driven innovation" was coined in studies carried out by the OECD.[1] For a number of years, the OECD worked on the basis of the OSLO Manual, which dealt with definitions of various types of innovation. Subsequently, the OECD began to address the combination of innovation and big data, which led the organization to focus on the combination of big data and innovation:

> The analysis of "big data," increasingly in real time, is driving knowledge and value creation across society, fostering new products, processes and markets; spurring entirely new business models; transforming most if not all sectors in OECD countries and partner economies.[2]

Clearly, the OECD is keen to expose the benefits of data-driven innovation at the macro level. Therefore, the organization has used a great deal of energy to investigate how data-driven innovation can improve productivity and efficiency in the public-welfare sectors of its member countries.

However, our interest is in data-driven innovation at the micro level, which we believe has been largely overlooked. This is regrettable as data-driven innovation represents a new, critical source of growth.

A disproportionately high amount of innovation takes place in the vibrant market for start-ups. Therefore, rather than looking to large companies' R&D departments in our search for innovation, we examine the collaborative relationship between startups and large corporations.

We believe that large corporations have been overvalued for quite some time, especially when it comes to complacent incumbents. Even though these corporations possess significant resources and scale, many of them are in desperate need of growth. In the wake of the COVID-19 crises, these characteristics become more relevant than ever as the traditional leaders seem to be hardest hit by the crisis. Some of them are also struggling to integrate the UN's Sustainable Development Goals (SDGs) into their business equations.

Conversely, startups seem to have been largely overlooked, at least by complacent incumbents. Startups often possess superior innovation capabilities and are, therefore, an important source of growth despite the fact that they often lack resources and scale.

As such, there is considerable complementarity between start-ups and large corporations. A natural starting point is to look into the possibilities of establishing collaborations between these two vastly different types of organizations. We do not intend to address all of the relevant subtopics of this collaborative relationship but rather just a very specific subcategory: how to appropriate value from digital platforms

Acquiring data-driven innovation

This topic is probably more important than ever in the wake of not only the recent general recession but also the subsequent COVID-19 crisis and SDG challenges. Growth can help large corporations and, on an aggregated level, society, conquer such issues.

This book deals with the more abstract history of innovation as well as the practical aspects of how one can acquire data-driven-innovation and appropriate value from it.

Our point of departure is our reflection on what innovation is in Chapter 2. The term "innovation" has become a colloquialism for almost anything that needs to be portrayed in a positive light. Who does not want to be innovative these days? A more thorough explanation of some of the key terms in this book, including "innovation," "data-driven," "artificial intelligence" (AI), and growth, is offered in Appendix D.

Human minds and human behavior are prone to systematic biases because humans are driven by experience and because they rationalize their behavior ex post. The fact that we also have "blind spots" adds to the challenge of innovation. The same is true for linear thinking, which is built into the human mindset. Linear thinking seems to contradict innovation, which often generates, for example, exponential (rather than linear) growth. In short, innovation essentially goes beyond our own comprehension.

These characteristics of the human mindset have several implications. First, we are often satisfied with business as usual—we are not intuitively minded for extreme growth or for the blitzscaling of companies. Second, we tend to focus on best practice and not next practice. In other words, our focus should be on how we can achieve something without precedence.

Chapter 3 moves us from the individual level to the company level. For many years, innovation was viewed as something that occurred behind closed doors in the R&D departments of large companies. However, this kind of closed innovation is ineffective in today's world. A typical example is an innovation process that takes place in closed circles over a considerable span of time and eventually fails to meet market requirements. Product-market fit, which is often believed to be perfect prior to the initiation of a product-innovation process, is simply not there when the rubber meets the road. To a considerable degree, this resembles the painful outcome seen when software developers in a closed environment use the "waterfall method" to develop new IT systems. When a new system is ready to go live after several years of development, the outside world has changed so much that the new system does not meet the demands of the receiving world.

For these reasons, a great deal of closed innovation has gradually been replaced with open innovation. Open innovation is characterized by cooperation with external parties. The basic philosophy of open innovation resembles the agile software innovation process in which changes and adjustments take place concurrently in order to arrive at the desired product-market fit. Such innovations can either take the market by surprise early on, which is known as a "black swan innovation,"[3] or they can gain a foothold initially and then move up the market. The latter is what Clayton Christensen famously coined "disruptive innovation,"[4] which refers to seemingly irrelevant innovations that subsequently constitute a major threat to leading firms.

Open innovation has been key for increasing the effectiveness of innovation but has done so at the expense of speed. The increase in effectiveness leads to innovations that better meet the requirements and needs of the market, especially consumers. However, opening innovation processes to include collaborative efforts with external parties such as universities, results in more time-consuming processes. The question then becomes one of how to kill two birds with one stone. In other words, can innovation processes be designed so that they are simultaneously both more effective and more efficient?

A new innovation philosophy is now emerging. Some label it a new generation of innovation and talk about innovation networks in which innovation is generated in conjunction with the "creative commons."[5] The idea is that a more open business model based on data-driven innovation provides broader access to the market, lowers the costs of innovation, and allows for risk sharing.

What do these ideas mean in practice? One implication is that it is now possible to buy access to innovation as a digital service. Big data, openness, digital media, and platform technologies are all trends that can be combined to support the advent of advanced platforms from which one can acquire and extract innovation. This creates a new approach to innovation. Should I build or buy? Invariably, a mix may occur, but the bottom line remains the same: Accessing innovation from which you can appropriate value through a digital platform represents an interesting path toward simultaneously ensuring both effectiveness and efficiency in terms of high speed and low costs. The collective term for this new opportunity is "data-driven innovation," which we will use for the remainder of this book.

Chapter 4 deals with data-driven innovation as the key to solving many of the challenges at hand. The chapter starts with an outline of the traditional build-or-buy decisions that are normal in many industries but are relatively new when it comes to innovation, which is characterized by a strong tradition of "building." The "buy" option is closely connected with the idea of bringing the difficult start-up/corporation collaboration to new heights through the new digital platforms that are quickly emerging.

A walk-through of these digital platforms is then presented along with a discussion of the extent to which the use of advanced AI has appeared. This culminates in a comparison of two of the leading platforms: CBInsights and Valuer.ai. Both platforms are based on the Software as a Service (SaaS) business model in which one can buy access to innovation through a subscription.

These new possibilities are addressed in detail in Chapter 5 where some of the functionalities are outlined (although some of the technicalities are discussed in the appendices). In addition, Chapter 5 includes some use cases of a "what if" nature, which examine the platforms that may be used under certain circumstances. In some cases, it may be prudent to rely on complementary input from more than one platform. In other cases, input from just one platform may suffice.

Chapter 6 brings us back to the combination of innovation and growth. Despite the fact that this book aims to illustrate the advantages of relying on data-driven innovation, we must remember that growth does not occur automatically, even if you manage to find the right needle in the haystack. The chapter introduces the term "exnovation" as an important condition for delivering growth and, thereby, appropriating value from data-driven innovation. Essentially, exnovation deals with scaling up valuable innovations through various means such as standardization, mass production, mass customization, and geographical expansion. In other words, it relates to ways of achieving market leadership, "black swan" effects, valuable disruption, first-mover advantages, and situational monopolies.

Some of the business cases presented in Chapter 6 illustrate that different combinations and chronologies lead to different outcomes. Nevertheless, we conclude that, ideally, innovation and exnovation should go hand in hand.

A book like this would normally close after Chapter 6, but our findings together with the world's hunger for growth inspire a reflection on whether a new digital modus operandi can serve as the vehicle for such growth. Therefore, Chapter 7 takes a look at the near future and a scenario in which data-driven innovation from digital platforms paves the way for the transformation of innovation processes from analogue to digital workflows. Such a transformation may result in the advent of what we call a *digital innovation economy*.

The sudden occurrence of the COVID-19 crisis in 2020 and the desire to address SDG challenges also gave rise to certain reflections, which are discussed in Chapter 7. These reflections relate not only to the wider repercussions of these two issues but also to the importance of accelerating services and solutions based on digital technologies and AI. Data-driven innovation may serve to mitigate the negative effects of such crises and help corporations and start-ups alike gain extra momentum.

Notes

1 OECD. (2015). *Data-Driven Innovation: Big Data for Growth and Well-Being*. OECD: Paris.
2 Ibid., p. 20.
3 "Black swan" is a term that was originally coined by Nassim Taleb (2010). *The Black Swan: The Impact of the Highly Improbable*. New York: Random House. It was used thereafter in the business strategy context, e.g., Andersen, M. M. and Poulfelt, F. (2014). *Beyond Strategy: The Impact of Next Generation Companies*. London and New York: Routledge.
4 See Christensen, McDonald, Altman & Palmer 2016 and Christensen, 2016 (same footnote as footnotes 3 and 4 in Chapter 3).
5 See Curley, M. and Salmelin, B. (2018). *Open Innovation 2.0: The New Mode of Digital Innovation for Prosperity and Sustainability*. Cham: Springer, p. 40f.

chapter 2

Human behavior
Dealing with innovation and the unknown

Innovation involves the creation and adoption of something new from which value can be appropriated. It might refer to new products, new features, or new ways of doing things. Innovation often creates unexpected outcomes, as it does not rely on rational and foreseeable decision-making. Instead, innovation involves the unthinkable—something believed to be beyond human imagination. Innovation identifies and harnesses the unknown and the unknown unknowns in the human mind. In contrast, common thinking is based on the known world. In common thinking, we recognize and deal with what we can see and comprehend as human beings, while innovation emerges in human blind spots. Innovation harnesses the unknown and represents what many would either have not thought of or regard as unthinkable. Invariably, it is challenging to turn unknowns into knowns, especially given humans' biases in our thinking. We create our own narratives, adopt a herd mentality, build on averages, and are overconfident in what we believe or think we know. Moreover, we are unable to acknowledge the full extent of our ignorance, and we are prone to reject dissonant information when presented with data that counteracts with our own beliefs.

"What I do not know I do not think I know either"

Innovation and the ways in which value is appropriated from innovation are the central themes of this book. When investigating innovation, one has to start by trying to understand the relevant dimensions of human behavior. This is because individuals carry out all innovation. While machines, buildings, and organizations might facilitate innovation, at the end of the day all innovation can be traced back to individuals and to interactions between individuals in teams and other kinds of collaboration.[1] Therefore, we need to take a deeper dive into the restrictions of the human mindset and the implications of those restrictions for human behavior.

In the modern understanding, the term "innovation" refers to something new. It originated from Latin, where *innovare* means "renewal" and *novus* means "new."

Human behavior

Innovation involves the creation and adoption of something new from which value can be appropriated to, for example, meet new requirements or unarticulated needs. As such, innovation essentially deals with something new and original. It might refer to new products, new features, or new ways of doing things. Innovations can appear in many forms such as technical innovations, organizational innovations, or business-model innovations.[2]

One of the first scholars to address the human mindset and innovation was Socrates. This theme is particularly evident in Socrates's discussions of knowledge. Socrates (470–399 BC) is widely quoted for what some label a paradox and others label a contradiction: "I know that I know nothing." However, more detailed studies of ancient Greek now show that this is in fact a quote taken from Cicero and has been wrongly attributed to Socrates, who should have been quoted along the following lines: "What I do not know I do not think I know either."[3] Socrates did not write his philosophical wisdom down—what we know of Socrates's teachings we know from his student Plato. Socrates is critical of people who think that they know something when, in fact, they do not. Therefore, the wiser path is to acknowledge and recognize what one does not know rather than to disregard this lack of knowledge.

At first glance, the difference between the two quotes may not be a big difference. However, a more distinct analysis reveals that Socrates, before 400 BC, instituted the first journey to establish a cognitive theory of knowledge and innovation. For the authors of this book, the correct quote given previously is key to answering the question of whether you can acquire or buy innovation. For Socrates, knowledge was the most valuable thing in life. However, he believed that it was better to seek knowledge than to claim knowledge that one does not really possess. This suggests that Socrates would probably have been a strong supporter of acquiring knowledge from external sources.

The incorrect quote – "I know that I know nothing" – is a contradiction, which fails to help us interpret how to orchestrate the combination of knowledge and innovation. The correct quote, therefore, is more useful in this regard.

"What I do not know I do not think I know either" is neither a paradox nor a straightforward contradiction. Socrates recognizes the severe limitations of human cognition in the phrase "What I do not know." This is in sharp contrast to the many innovators and key employees of large corporations who proclaim that they know better than anybody else. The wording "I do not think" is a wise and conscious declaration – possibly a comment on the world of the unknown and, undoubtedly, a comment on Socrates's own role as a human being with limited cognition. "I do not think I know either" implies the recognition of blind spots from a very conscious and, occasionally, highly provocative Socrates.

This is also evident in Socrates's retelling of the story of Thamus, the Egyptian king, and Theuth, the inventor of the written word. Socrates explains that Theuth presented his new invention, "writing," to King Thamus. Theuth tells Thamus that his new invention "will improve both the wisdom and memory of the Egyptians." King Thamus is skeptical of the new invention and rejects it, arguing that the written word would infect the Egyptian people with fake knowledge as they would be able to gather information from external sources and no longer be forced to retain large quantities of knowledge themselves.[4] This is a telling story – Socrates stood in stark contrast to King Thamus and, in many ways, became the first ambassador for the importance of gaining knowledge from external sources.

Collectively, this resembles the mindset of a philosopher who is keenly aware of not only his own cognitive limits but also the fact that there is something beyond his own cognition. Rephrased, this could sound like: "I cannot know what I do not know." Socrates, in his own way, recognizes that innovation requires some kind of cross-boundary external assistance.

Human emotions, biases, and delusions

Human psychology and studies of the human brain indicate that we are prone to creating false narratives that lead us to believe that we are in control of things when in fact we are not. In his groundbreaking studies, Jonathan Gifford describes this as follows:[5]

> Our choices, it turns out, are driven by older instincts, emotions and drives, while what we take to be our 'self' gaily rationalises what we have done after the event.

Gifford provides several examples of how we can be blindsided by various phenomena such as stock-market bubbles, the dot-com bubble, epidemics, the collapse of large corporations, or a credit crunch. He concludes as follows:[6]

> What we experience as rational decision-making is driven by impulses and urges of which we are barely aware: an overwhelming urge to take what is available now before someone else gets it; an inbuilt tendency to follow the herd; a strong sense of what is equitable and fair; an instinct to trust our own group and to be hostile to outsiders. . . . These unexamined motivations are what cause us to be blindsided.

What Gifford essentially points out is that we are more driven by emotions than we may realize, and that the rational component of our thinking often only emerges when we attempt to retrospectively justify and rationalize events. In all other contexts, human beings are selfish, greedy, shortsighted, and prone to mass delusion.

This is also in line with the thinking of Nobel laureate Daniel Kahneman, who studies behavioral microeconomics. Kahneman states that "inconsistency is built into the design of our minds,"[7] "our minds are susceptible to systematic errors,"[8] and "overconfidence is fed by the illusory certainty of hindsight."[9]

In many cases, individuals are either unaware or not fully aware of these mechanisms and, therefore, unable to combat biases in their thinking. Consequently, a halo effect emerges with the aim of rationalizing mediocre innovations ex post. With regard to innovation in the future, people make their stipulations based on what they are able to observe here and now.

This is also obvious from a study of the literature on strategic management where promotional titles such as *In Search of Excellence*, *Built to Last*, *From Good to Great*, and *What Really Works* indicate that a successful formula for running a company has

Human behavior

Table 2.1 Delusions related to the halo effect

The halo effect	The tendency for the performance analysis of a company to reflect only the overall results
The delusion of correlation and causality	*Do we always know which thing causes what?*
The delusion of single explanations	*There is rarely one specific factor that can explain everything.*
The delusion of connecting the right dots	*It is impossible to isolate the reasons for being successful, as most studies do not compare such organizations with less successful companies.*
The delusion of rigorous research	*The data may lack the required quality.*
The delusion of lasting success	*Sustainable formulae for success do not exist.*
The delusion of absolute performance	*Company performance is always relative.*
The delusion of the wrong end of the stick	*It may be true that successful companies pursue a focused strategy, but this does not mean that a focused strategy always leads to success.*
The delusion of organizational physics	*Company performance does not have the certainty of an immutable law of nature, so it cannot be predicted with the accuracy of the natural sciences.*

been found. Essentially, these best-selling books suffer from a variety of business-related delusions of which the halo effect is the most important.[10]

The research behind the delusions listed in Table 2.1 highlights severe biases among company executives in relation to strategic management issues. The "halo effect" refers to a cognitive bias found among executives who will gladly attribute successful outcomes to their own performance and, conversely, hide business failures or even redefine them as successes. This is similar to the Dunning-Kruger effect – being ignorant of one's own ignorance and, consequently, grossly overestimating one's own skills and capabilities.[11]

With a focus on rebel ideas, Matthew Syed talks about "perspective blindness" as follows:

> We are oblivious to our own blind spots. We perceive and interpret the world through frames of reference but we do not see the frames of reference themselves. This, in turn, means that we tend to underestimate the extent to which we can learn from people with different points of view.[12]

He even goes a step further by suggesting that human nature itself means that humans are likely to not only share each other's blind spots but also to reinforce them. He terms this "mirroring." When you are surrounded by people who reflect your picture of reality and whose picture you reflect back to them, it is easy to become overconfident in judgements and assessments that are factually incomplete or simply wrong.[13]

With emotions, overconfidence, bias, delusions, and mirroring operating within most humans, it is no surprise that innovation and outlier experiences encounter difficulties from time to time.

Linear thinking and attempts to remove outliers

A straight line between two points is the most efficient way to get from one place to another. In strict terms, this is linearity. Linear thinking is the process by which "linear thinkers" put things in order as they experience them. This thinking process proceeds in a sequential manner, like a straight line.[14]

Statistically, this has interesting consequences for how you crunch your data when dealing with innovation and predictions for the future. In a world increasingly characterized by data-driven features, this is particularly important. In line with what is often taught at business schools in relation to market economics, the average is important both because the Gauss curve displays many observations around the mean and because you often wish to level out the rest of your observations, especially those that are too far from the mean. When it comes to causality, the average is also important in linear cause-effect thinking, as "regression to the mean" becomes dominant, often in order to strongly emphasize an argument about causality. This methodology helps achieve a nice correlation and allows you to focus on expected observations and leave out unexpected observations.

Arguably, the focus on the average is not particularly helpful when it comes to innovation. Innovation is rarely, if ever, average. Therefore, average is the enemy when it comes to innovation – something truly new is by no means "average."

With the phrase "birds of a feather flock together," Plato addressed the danger of intellectual conformity in his masterpiece *The Republic*.[15] The dangers of homophily were a preoccupation in Ancient Greek culture. Plato's student Aristotle, writes that people "love those who are like themselves."[16] Similarly, some of the leading scientists from the earliest days of Western civilization acknowledged the comfort of being surrounded by people who mirror your own perspectives and hold similar prejudices. One could call this a kind of self-validation, and brain scans indicate that this stimulates the pleasure centers in the human brain.

This is a major concern in the context of innovation, as it fosters clone-like attitudes and herd-like behavior. Cloning is basically the opposite of innovation. Thus, the pessimistic conclusion based on human cognitive findings is that it is difficult to generate innovation in a world of "average," unless certain means of measurement other than "average" are taken into account.

Innovation often appears as a surprise in the ecosystem. As development processes evolve, innovative ideas may remain in hibernation in an unknown start-up context. When they become known and utilized, the ecosystem is taken by surprise. Given the fact that we have a systematic human bias toward something, which is not surprising but average, this is an important characteristic when innovative ideas reach either the corporate garage or the market.

Innovation often creates unexpected outcomes because it does not rely on rational and foreseeable decision-making. Instead, innovation involves the unthinkable – something believed to be beyond human imagination. Innovation identifies and harnesses the unknown and the unknown unknowns in the human mind. In contrast, linear thinking is based on the known world. In other words, in linear thinking, we recognize and deal with what we can see and comprehend as human beings, while innovation emerges in human blind spots. Innovation harnesses the (previously) unknown and represents what many would either have not thought of or regarded as unthinkable. This move into the world of knowns versus unknowns is central to the understanding of the human mind and its inherent restrictions.

Invariably, it is challenging to turn unknowns into knowns especially given the humans' biases in our thinking. We create our own narratives, think linearly, adopt a herd mentality, build on averages, and are overconfident in what we believe or think we know. Moreover, we are unable to acknowledge the full extent of our ignorance, and we are prone to reject dissonant information when presented with data that counter our own beliefs.

It is, therefore, time to return once again to Socrates: "What I do not know I do not think I know either."

We expand on this idea in the following section, which deals with the knowns and the unknowns.

What do we know and what is unknown?

We will now try to put ourselves in the shoes of a corporate decision-maker such as a CTO in an innovation department in a large, blue-chip company, or a distinguished partner in a venture capital (VC) company. In particular, we will attempt to take this decision-maker's mindset into consideration.

The starting point is to pair this individual's mindset (acknowledged innovation = "known") and unacknowledged innovation (= "unknown") with the external knowledge on innovation that is generally available.[17] We make this distinction between what is known to the world (external recognition) and what is known to the decision-maker (own mindset) to derive the two-by-two matrix shown in Table 2.2.

The nature of innovation differs depending on where we are in the two-by-two matrix, as illustrated in Figure 2.1, which builds on the four quadrants shown in Table 2.2. In Figure 2.1, we list the types of innovation that are typical for the four quadrants (e.g., blind spots in quadrant 2).

The lower dotted line in Figure 2.1 represents the typical limits of traditional thinking. The more traditional thinking occupies a large chunk of Q1 in Figure 2.1 – the known-known quadrant and, at best, a slice of the "opportunity spot" (i.e., what is known to you but unknown externally). This is on par with general human thinking as discussed earlier in this chapter.

Table 2.2 Knowns and unknowns

		External Recognition	
		Known	Unknown
Own Mindset	**Unknown**	Quadrant 2 / Q2	Quadrant 3 / Q3
	Known	Quadrant 1 / Q1	Quadrant 4 / Q4

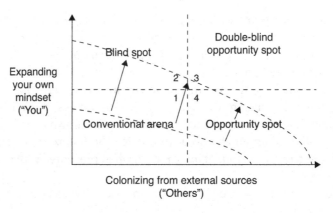

Figure 2.1 Improving the mindset for the innovation journey

The type of innovation in Q1 is typically what takes place in the R&D departments of large corporations. This is where one can find views along the lines of "we are capable of doing things ourselves," "internal development is better than anything else," and "my experience tells me. In other words, this is the conventional type of innovation.

Cognitively, beyond the fact that we are only dealing with readily available innovations, we are dealing with the limitations tied to the human tendency to engage in linear thinking, reject dissonant information, and create our own narratives. Notwithstanding the fact that the overwhelming part of innovation and innovation processes can be categorized as belonging to Q1, it is cognitively dissatisfactory to only rely on this type of innovation. Today, the risk of missing the target and the risk of failure when only relying on the known are simply too high. The arrows in Figure 2.1 illustrate the idea of expanding and improving one's own mindset in order to be better positioned for the innovation journey. The figure clearly shows the improvements that data-driven innovation for digital platforms can generate. One obvious example is to find innovation by accessing the new digital platforms, thereby learning about innovations that might otherwise remain unknown.

In the search for methodologies and tools to make the unknowns known, we need to actively work on avoiding biases in our thinking. As a starting point, innovators have to avoid linear thinking. This results in a shift away from a focus on the "average" or a "regression to the mean" to a focus on outliers, exceptional cases, and unprecedented types of innovation. The presence of a self-contained innovation mindset that encompasses the "not invented here" (NIH) syndrome might lead you to reject innovative ideas.

Even the renowned Microsoft CEO Steve Ballmer missed this point when he addressed the iPhone's lack of a physical keyboard (note his Q1 thinking):

> 500 dollars? Fully subsidized? With a plan? I said that is the most expensive phone in the world. And it doesn't appeal to business customers because it doesn't have a keyboard. Which makes it not a very good email machine. . . . Right now, we're

Human behavior

selling millions and millions and millions of phones a year. Apple is selling zero phones a year. In six months, they'll have the most expensive phone by far ever in the marketplace.[18]

Apple, which orchestrated this innovation in a start-up environment, relied on non-linear thinking. It deliberately made waves with the iPhone in 2007 by introducing a range of innovative features, services, and designs as well as unprecedented marketing and sales tactics.

None of the most recent market successes – Google, Facebook, LinkedIn, Airbnb, Alibaba, Tencent, and Baidu – buy into the linear mindset. Not only do they provide innovative and appealing offerings, but their market approach is also clearly nonlinear. This is evident in their extreme focus on exponential growth and their prioritization of speed. For example, Google used an innovative and previously unseen approach to achieve growth by entering into a partnership with AOL in its early days. Google had revenue of just US$19 million in 2001. It announced a surprising and unexpected deal with AOL in May 2002. Of the revenue the partnership would generate, 85% would go to AOL. Retroactively, this seems to have been the right decision as Google's revenues increased to an impressive US$347 million in 2003 despite the bursting of the IT bubble.

A new term for such growth hacking with subsequent immediate and exponential growth in revenue was coined by Reid Hoffman and Chris Yeh: "blitzscaling." The terms build on human thinking and looking beyond the Q1 type of mindset. A more elaborate description of "blitzscaling" is provided in Appendix A.

Similarly, Amazon has engaged in substantial blitzscaling on several occasions. One example is shown in Table 2.3, with Amazon delivering growth of 322x over three years.

NIH syndrome does not exist in this context as the focus is on extreme and rapid growth. Instead of rejecting radical and external ideas, some companies promote and prioritize these ideas. A quote from a study on companies based in Silicon Valley serves to illustrate this point:

> We talked to hundreds of entrepreneurs and CEOs, including those of the world's most valuable companies, such as Facebook, Alphabet (Google), Netflix, Dropbox, Twitter, and Airbnb. . . . Even though the stories of their companies' rise were very different in many ways, the one thing they all had in common was an extreme, unwieldy, risky, inefficient, do-or-die approach to growth.[19]

Table 2.3 Amazon's blitzscaling from 1996 to 1999[1]

	Number of employees	Turnover, USD
1996	151	5.1 million
1999	7,600	1.64 billion
Increase	50x	322x

1 Hoffman, Reid and Yeh, Chris (2018). *Blitzscaling. The lightning-fast path to building massively valuable companies*, Currency, New York, p. 23.

Notice that some of the human characteristics here – risky, inefficient, a do-or-die approach – are far from the NIH syndrome.

Moreover, the acquisition of innovative companies may be part of the growth story if things do not scale up quickly enough through organic growth, and "internal" innovation does not sufficiently deliver with regard to business development. Google serves as a case in point. Somewhat contrary to its general image, Google has grown considerably by successfully acquiring innovative companies such as DeepMind, Nest, YouTube, and Android.

Making unknowns known

When trying to make unknowns more known, we need to take a closer look at quadrants 2, 3, and 4. In order to delve deeper into quadrant 2, the starting point is to define it as a blind spot. Knowledge is not recognized individually or collectively. This means, for example, that the R&D department of a large corporation may be blind owing to the aggregation of individual blindness, relating to the Dunning-Kruger effect and the herd tendency. However, relevant innovation is readily available outside the organization – that innovation needs to be acquired. In order to successfully extract value from external sources, an openness to experimentation is necessary. Trial and error is important for achieving exponential growth through innovation.[20]

The acquisition of innovation from the outside helps managers enlarge their known universe and make former blind spots visible. This, in turn, expands the active innovation arena.

Quadrant 4 deals with innovation of which you are aware that is not readily available outside the organization. For instance, when Steve Jobs and his team at Apple invented the iPhone, strict security precautions were implemented to keep the world unaware of the innovation.[21] However, there may be dormant innovation out there that has not yet been recognized or of which human consciousness is unaware. By probing at this level, we can take advantage of the opportunity territory in Q4 relatively easily.

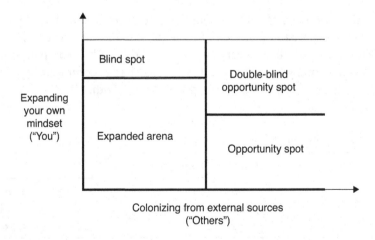

Figure 2.2 Expanding from known innovation (Q1) in order to capitalize on innovation from blind spots (Q2)

Human behavior

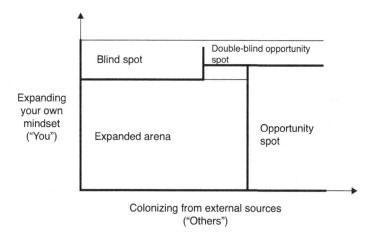

Figure 2.3 Expanding from known innovation (Q1) in order to capitalize on innovation from opportunity spots (Q4)

The move to quadrant 3 is more difficult. This quadrant, which we label the "double-blind opportunity spot," gained considerable attention when former US Secretary of Defense Donald Rumsfeld pointed out: "There are also unknown unknowns. There are things we don't know we don't know."[22]

Although this is the most difficult quadrant to work with, one should attempt to develop a cognitive understanding of the unknowns. Expansion into this arena may be achieved by mentally preparing to acquire innovation from external sources, as we also saw when discussing Q2. However, there is an important cognitive difference between the approaches to Q2 and Q3. Q2 deals with innovation of which you are unaware can be acquired by external sources. This requires the R&D manager to submit an open request to capable external sources, such as: "Please supply us with innovation that complements the innovations emerging from our R&D department." However, how can humans address the unknown unknowns when the innovation is not even known to competent, external sources?

The answer to this tricky question relates to new techniques that go beyond the comprehension of the human brain. A number of corporations have successfully opened the door to double-blind opportunity spots by utilizing special analytics based on big data, AI, and sophisticated algorithms. A fairly comprehensive global database of start-ups representing various types of innovation may be the starting point, as we discuss in Chapters 4, 5, and 6. By analyzing big data on innovation through the lenses of AI and advanced algorithms, unknown unknowns can be transformed into the end-destination of unprecedented innovation from which value can be appropriated.[23] One approach is to transform the unknown unknowns into knowns in two stages. The first stage is to transform this valuable double-blind opportunity innovation into unknown knowns (Q2). In the second stage, organizations can remove their blind spots and convert the unprecedented innovations into fully known knowns through presentations and cognitive preparedness.

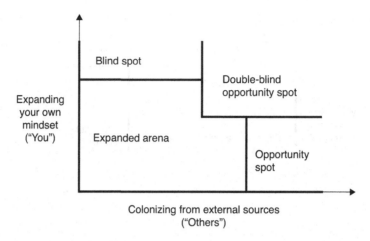

Figure 2.4 Expanding from known innovation (Q1) in order to capitalize on innovation from double-blind opportunity spots (Q3)

Getting around human biases and harnessing innovation

The human factors operating against innovation, especially innovation related to unknowns, are rooted in Socrates's statement: "What I do not know I do not think I know either." Given this statement, it is only natural to take a deep dive into the world of the unknown. However, it is somewhat against human nature to recognize that there are major loopholes in our knowledge system.

The main challenges regarding human biases when it comes to innovation are systematic errors, being blindsided, the halo effect, and the Dunning-Kruger effect. These factors produce risk-averse, linear thinking, which makes it difficult and sometimes impossible to take a deep dive into the world of the unknown.

The inherent human bias tied to innovation is only one side of the coin. The other is to overcome the challenges by, for example, working systematically to make unknowns known by adopting Socrates's high level of consciousness and executing accordingly.

As a key tenet in this book, it is worthwhile to quote John Chambers of Cisco:[24] "Companies that don't partner with start-ups will become irrelevant." Why would a seasoned CEO from an innovative company make this statement? Interestingly, Cisco was at the top of company valuations in the year 2000, but it later lost this position, largely because other companies blitzscaled and grew exponentially. Traditionally, Cisco's acquisition activities focused on well-established companies. However, start-ups are in an earlier stage, which creates a number of advantages for corporations.

Many start-ups, not all, are inventing something new – similar to "unknown innovation." If corporations can gain access to an innovation through a start-up, they can tap into the world of unknowns.

Companies like Nokia miss opportunities to obtain innovative input from the outside, owing to their self-contained, overconfident cultures. Executives in these companies do not perceive a need for external input, and they fail to search for the unknowns

Table 2.4 Summarizing key points from "as is" and "unknown innovation"

"As is" – business as usual	"Unknown innovation"
Creation of something new	Creation and adoption of a new unknown so that value can be appropriated
"What we have found out" (the knowns)	"Something beyond our comprehension" (the unknowns)
Linear	Exponential
Regression to the mean/extinction	Focus on outlier experiences
The halo effect	Reflections/acceptance
Business as usual	Blitzscale
Best practice	Next practice

because they believe that they are the "champions" within their field – "*if some innovation was really worthwhile, we would have known about it.*"

Even Google, which is highly regarded for its internal innovation efforts, acquires input from start-ups. Furthermore, such innovations have been instrumental in Google's acquisition of a number of companies. Despite the general sentiment to the contrary, Google has derived far more innovative power from external sources than from its internal efforts. In fact, its internal efforts have led to a number of failures, including Google Glass, Google+, Google Lively, Google Wave, Google Buzz, Allo, Google, and Google Nexus One. On the contrary, Google's acquisitions of start-up innovations have been very successful.

In contrast to Nokia, Google's performance relies on internally orchestrated innovation as well as innovation acquired from start-ups. With John Chambers's statement in mind, it seems as if Google's strategy is more viable. We therefore focus on start-up-innovation in the following chapters.

Before beginning to expand on ways of avoiding human biases when dealing with innovation, it might be worthwhile to sum up the main finding in this chapter on making the unknowns known when dealing with innovation. This is precisely the focus of Table 2.4 at a generic level, while a detailed discussion of exploration is left to the following chapters.

This brings us to Chapter 3, where we will examine how these individual biases play out at the company level.

Key reflections

A   Innovation is the focal point of this book. Consequently, it is necessary to start by defining innovation. Notably, more than 40 definitions exist. In our definition, we stress that from a business perspective, the issue is not just the identification of innovation but also the adoption of innovation. Without adoption, no value is created, and value creation is key. *How do you define innovation? What is the difference between the colloquial meaning of innovation and business-related definitions and applications of innovation?*

B   In this chapter, we dealt with a number of restrictions in the human mindset, such as overconfidence, the tendency to follow the herd, delusions resulting from the halo effect, linear thinking, and the rejection of outliers. These characteristics may severely restrict seeking innovation from the broader world of the unknown and result in a tendency to only look for something nearby in the (well-)known universe. *What kind of restrictions are most important when it comes to creating and adopting value from innovation? What are the paths to finding innovation in the world of the unknown?*

C   In this chapter, we discuss the case of Google. Google has engaged in a considerable amount of proprietary innovation that has culminated in failed business ventures, such as Google Glass, Google+, Google Lively, Google Wave, Google Buzz, Allo, Google, and Google Nexus One. At the same time, Google has acquired numerous start-ups with strong innovative capabilities, and it has appropriated significant value from several including DeepMind, Nest, YouTube, and Android. *Can the failure of in-house innovation and the success of acquired innovation be explained by the nature of the human mindset? What does it take to move from best practice to next practice?*

Notes

1 Foss, N. and Pedersen, T. (2019). Microfoundations in International Management Research: The Case of Knowledge Sharing in Multinational Corporations. *Journal of International Business Studies*, 50: 1594–1621.

2 Foss, N., Pedersen, T., Pyndt, J. and Schultz, M. (2012). *Innovating Organization & Management: New Sources of Competitive Advantage*. Cambridge: Cambridge University Press.

3 Postman, N. (1992). *Technopoly*. New York: Vintage, p. 74, and Plato (2012). *Apology*, section 21d in *Plato in Twelve Volumes*. Translated by Harold North Fowler. Introduction by W. R. M. Lamb. Cambridge, MA/London: Harvard University Press/William Heinemann Ltd.

4 Plato (1966), *Phaedo*, by Plato, full text (English & Greek). *Plato in Twelve Volumes*. Translated by Harold North Fowler. Introduction by W. R. M. Lamb. Cambridge, MA/London: Harvard University Press/William Heinemann Ltd, pp. 274e–275b.

5 Gifford, J. (2012). *Blindsided: How Business and Society Are Shaped by Our Irrational and Unpredictable Behaviour*. London: Marshall Cavendish Business, p. 9.

6 Ibid., p. 222f.

7 Kahneman, D. (2011). *Thinking Fast and Slow*. New York: Allan Lane, p. 385.

8 Ibid., p. 16.

9 Note 13: Ibid., p. 14.

10 Cf. Rosenzweig, P. (2007). *The Halo Effect . . . and the Eight Other Business Delusions that Deceive Managers*. New York: Free Press.

11 Cf. Dunning, D. (2011). The Dunning-Kruger-Effect: On Being Ignorant of One's Own Ignorance. *Advances in Experimental Social Psychology*, 44: 247–296.

12 Syed, M. (2019). *Rebel Ideas: The Power of Diverse Thinking*. London: John Murray, p. 20

13 Ibid., p. 24.

14 Linear thinking is illustrated by a joke. Two people are sitting next to each other in a plane when they observe there is a fire in one of the two engines. The pilot informs everybody there is no need to worry, but they will be delayed by an hour. A little later, the two individuals observe that there is also fire in the other engine. One says to the other: "What a pity. Now we will be two hours late."

15 Although whether this translation is correct is disputed, phrases.org.uk states: "Plato's text can be translated in other ways and it is safe to say it was Jowett in 1856, not Plato in 380BC, who considered the phrase to be old. The lack of any citation of it in English prior to the 16th century does tend to suggest that its literal translation wasn't present in The Republic – a text that was widely read by English scholars of the classics well before the 16th century" (www.phrases.org.uk/meanings/birds-of-a-feather-flock-together.html).

16 In his *Nicomachean Ethics*.
17 This is a further development of what appeared in Appendix 1. Andersen, M. M. and Poulfelt, F. (2014). *Beyond Strategy. The Impact of Next Generation Companies*. New York: Routledge.
18 Cf: www.macrumors.com/2016/11/07/former-microsoft-ceo-steve-ballmer-wrong-iphone/.
19 Ibid., p. 14.
20 Ibid.
21 Lashinsky, A. (2012). *Inside Apple: The Secrets Behind the Past and Future Success of Steve Jobs's Iconic Brand*. London: John Murray.
22 Cf. www.brainyquote.com.
23 Cf. some of the case descriptions in Andersen, M. M. and Pedersen, T., *From Innovation to Exnovation: How Data-driven Innovation Platforms Change the Game*, cf. the section under the heading: "Blitzscaling Exnovation on Top of Innovation", *CBS* 2020.
24 Cf. www.bloombergquint.com.

chapter 3

Firms and innovation
Sources of innovation

Since the industrial revolution, the main model of innovation has been the "closed" approach, which focuses on internal development and protection of intellectual property (IP) rights. However, this inward-looking model has some serious shortcomings as the firm bears the entire cost and risk of product development. It is simply becoming more of a risk to ignore all the innovative ideas created outside the company's boundaries. Therefore, more firms are adapting open innovation, which is grounded in the recognition that firms can harness multiple sources of knowledge to enhance their own innovation. When relying on an open innovation model, a company does not strive to generate the best ideas on its own. Rather, it seeks to more effectively utilize internal and external ideas in order to arrive at more valuable types of innovation. Open innovation has primarily focused on collaboration with the closed counterparts, but with data-driven innovation, new possibilities emerge for searching on a global scale for innovative and disruptive ideas and tapping into the world of the unknowns.

Open versus closed innovation

For many firms, the exploitation of new and innovative ideas is a matter of life or death in the long run. However, the blindness in innovation behavior that we see among individuals aggregates to the firm level. These same biases may be reflected in how firms source new knowledge and innovation.

Views on sources of innovation have changed a great deal over time.[1] Since the industrial revolution, the main model of innovation has been focused on internal development and protection of intellectual property (IP) rights through, for instance, patents with the aim of creating a competitive advantage for the innovating firm. This is a characteristic of the traditional "closed" approach to innovation. Closed innovation rests on a number of assumptions including the need for a firm to:

1   Discover, develop, and market the innovation on its own;
2   Recruit and rely on internal talent to deliver innovation; and

3   Adopt restrictive IP management practices in order to prevent other firms from benefitting from the innovation.

The key aspect of the closed approach is that firms want to control and own the knowledge in order to appropriate the value of the innovation on a strictly proprietary basis. They essentially handle everything themselves within the boundaries of the firm. In other words, the firm "owns" the entire innovation value chain, from generating a new idea to developing it into a new offering for customers. This is in line with the biases and inward-looking view on innovation evident among individuals, which is not surprising as the firm level represents an aggregation of interactions among individual employees.

Some characteristic statements from a large corporation with a closed approach to innovation may include:

> We have hired very skilled professionals and we have been in this business longer than anyone else. Therefore, we are better able to innovate.

> When we put all of our R&D resources together, we are unbeatable. We have so many skilled PhDs and other resources at the highest possible level, so when our teams get together, the concerted effort of their activities is the best in the world.

Employees are also actors in the closed approach. As such, they are often an obstacle to open innovation even if the company's leadership supports open innovation. Statements along the lines of "I am very experienced, I know the ins and outs of this industry and therefore, I do not need input from the outside" are not unusual. The fear of being replaced or being made redundant often leads to such irrationality among employees, as Machiavelli suggested some 500 years ago.

This approach may have had some value in a world characterized by a low degree of globalization with a higher degree of certainty, and the absence of digitalization and machine learning, AI, and big data. In such a slow-moving world, some firms could afford to take the time to develop everything themselves. However, even then, it might have been dangerous to assume that the firm itself would be better than anyone else at generating all of the knowledge needed for innovation.

While the closed approach to innovation might have served firms well in the past, this inward-looking model has some serious shortcomings. First and foremost, the firm bears the entire cost and risk of product development in this approach. This is becoming increasingly untenable as offerings become more complex and globalization progresses. Companies are becoming more specialized, and knowledge is increasingly scattered around the globe. Therefore, firms can no longer assume that they can obtain the most innovative knowledge from their own pond. It is simply too risky to ignore the innovative ideas that are being developed outside the company's boundaries. It is also important to note all of the restrictions in the human mindset and behavior discussed in Chapter 2, which have significant effects at the firm level.

As a mitigating response, more firms are adapting open innovation, which is grounded in the recognition that firms can harness multiple sources of knowledge to enhance their own innovation. In other words, when relying on an open innovation model, a company does not strive to generate the best ideas on its own. Sources of knowledge and innovation typically include suppliers, research centers, universities, customers, competitors, and companies with complementary offerings.[2] In contrast to the focus on exclusivity and ownership that is associated with the closed approach, the essential characteristics of the open approach are access to, use of, and application of knowledge for innovation.

The initial idea of open innovation was based on the observation that many innovative companies deviated from traditional practices (Chesbrough, 2003).[3] Since the emergence of the open innovation concept in the early 2000s, companies have faced the challenge of embracing new forms of innovation, especially open communities they do not have direct control over. Open innovation requires collaboration amongst distributed but interdependent actors who rely on each other's capabilities for value creation and capture. The benefits of the open innovation approach relate to the increased reach of knowledge and innovation, and the sharing of the risk of innovation with external partners.[4]

The main differences between the closed and open approach to innovation are listed in Table 3.1.

One limitation of open innovation discussed in the literature, is that openness typically relates to existing collaborators such as suppliers and customers. Therefore, although open innovation is accompanied by a broader reach of knowledge and innovation, the firm is not operating in the unknown universe (see Chapter 2). The reach is typically limited to known collaborators or the closed network of existing partners, while more distant sources of knowledge and innovation are rarely considered. However, some attempts at crowdsourcing, where companies try to engage a broad range of innovators regardless of their location, are emerging. Notably, crowdsourcing was not part of the original definition of open innovation. Therefore, a new modality of innovation is coming to the fore.

Table 3.1 Main characteristics of open and closed innovation

	Closed innovation	Open innovation
Innovation model	Innovations emerge from the company's internal resources	Use of external knowledge to improve and accelerate the company's own innovations
Source of ideas	Internal	Exchange of ideas beyond firm boundaries
Intellectual property	Own know-how is treated confidentially in order to protect it	Innovation does not have to be legally protected in order to profit from it
Role of suppliers and customers	Passive recipients	Active co-innovators
Competition	Winner is the first actor to bring the innovation to the market	Winner is the actor that offers the best use of the innovation
Costs/benefits	Comparatively high degree of efficiency (speed), low degree of effectiveness	Comparatively high degree of effectiveness, low degree of efficiency (speed and costs)

Firms and innovation

This trend has recently been strengthened as digitalization and machine learning offer opportunities to expand the reach of and search for knowledge and innovation. In fact, the reach can be expanded to the entire world and not just the existing network. With the spread of the internet and all of the information it makes readily available, the costs of communication and of accessing that information have been sharply reduced. Moreover, the development of AI has made it easier for firms to identify, realize, and evaluate new and innovative knowledge from around the world. Accordingly, some larger firms rely on innovation intermediaries, idea scouts, and AI-based firms to seek, screen, and curate innovation, which they then review for commercial potential. We denote this emerging trend in which the whole world serves as the pond for seeking knowledge and innovation as "data-driven innovation." As data-driven innovation is the core topic of this book we shall outline this new way of gaining access to innovation in detail in Chapters 4, 5, and 6.

There is also a growing need to find a new modality of innovation creation and adoption, and new ways to increase efficiency and effectiveness in this regard.[5] As illustrated in Table 3.1, traditional closed innovation is increasingly inadequate when it comes to effectiveness. A classic example of low effectiveness is that innovation developed in a closed environment often does not fit the market. Therefore, it falls into the failure category. Open innovation is widely recognized as generating greater effectiveness and better product-market fit. However, this comes at the expense of efficiency as open innovation may take longer to produce and incur more costs. For example, one of the fastest ways to obtain access to some kind of open innovation is to rely on accelerator programs. An accelerator program alone lasts at least one year on average, while you can gain access to even better data-driven innovation from digital platforms in just a matter of weeks.

Consequently, there is a growing need for a modality in which the trade-offs between efficiency and effectiveness can be removed. Data-driven innovation seems to be important in this regard for a number of reasons:

- Data-driven innovation from some of the new platform vendors allows organizations to tap into the world of the unknown and, thereby, find the innovative "needle in a haystack," or they can compare innovation to cohorts of similar types of innovation on a global basis. *As such, data-driven innovation can dramatically increase the effectiveness of innovation and, thereby, achieve better product-market fit.*
- Data-driven innovation means that search missions can be defined instantly. Moreover, in some instances, results can be delivered in real time and, in other instances, can be provided on a predefined regular basis (e.g., monthly). Data-driven innovation is generally cheaper than getting a development team up and running. *As such, data-driven innovation generally saves both time and costs, thereby dramatically increasing the efficiency of innovation when compared to the standalone modalities of closed and open innovation.*
- Data-driven innovation from digital platforms constitutes and supports a digital workflow. This digital workflow serves as a substitute for traditional analogue workflows, which are characterized by negative externalities (e.g., highly time consuming, likelihood of human error, the need for physical presence and travel). *As such, data-driven innovation paves the way for the digitalization of innovation workflows.*

In addition to high efficiency and high effectiveness, several other characteristics of data-driven innovation deserve to be highlighted, given the human restrictions with regard to innovation. As we discussed in Chapter 2, these restrictions relate to characteristics including emotions, the tendency to follow the herd, and overconfidence.

Much data-driven innovation is generated by machine learning and more advanced types of AI. Dealing with AI on a digital platform means that there are no emotions attached – the AI innovation mission is completed without consideration for emotions, personal preferences, or idiosyncrasies related to, for example, nationalities, specific technologies, or certain types of products and services. In the absence of data-driven innovation, R&D staff members will often be affected by personal preferences, emotions, and overconfidence, as they are human beings. The AI system delivers on its assignment without any involvement of sentiments and subjectivity.

Given these attributes, we narrow the focal issue of this book to data-driven innovation from a selection of the new digital platforms.

Disruptive innovation

In the modern world, the degree of uncertainty is increasing at an accelerated pace, and the innovation scene is no longer just local or regional, but global in nature. As a result, businesses can no longer rely solely on forecasting and planning mechanisms. Managers should expect business models and patterns to change and, more importantly, need to implement this change themselves.

One term frequently associated with these dynamics in the business world is "disruptive innovation." Clayton Christensen developed his original framework on disruption in 1997 with the objective of explaining why incumbent firms that are leaders in their respective industries regularly fail when confronted with change (Christensen, 1997).[6] A starting point in understanding the theory of disruption is the difference between sustaining and disruptive innovations. Christensen describes sustaining innovations as those that improve existing products' performance with the aim of catering to the wants and needs of established customers. While these innovations can be incremental, such as a move from the iPhone 6 to the iPhone 7, or radical, such as the development of the first iPhone, they all enable companies to sell more of their products to their existing and profitable customer base. These innovations are of value to the firm's mainstream consumers in its major, high-performing markets. Historically, leading incumbent firms have almost always been successful in establishing sustaining innovations.

In contrast, disruptive innovations bring a completely new value proposition to the market. At the time of their launch, mainstream customers perceive these innovations as inferior. Generally, disruptive innovations can be described as cheaper and simpler, but they are often also more convenient to use than established products. New entrants lead in the development and commercialization of disruptive innovations, while incumbents frequently fail because of them.

How do disruptive innovations emerge, and why do they constitute a major threat to incumbent businesses? Disruptive innovations do not target mainstream markets. They are launched, usually by entrant firms, as new products or technologies.

Firms and innovation

At the time of their launch, mainstream consumers are not compelled to purchase them. According to Christensen, disruptive innovations can either address the low end of a market, which leading incumbents view as unattractive, or an entirely new market. A prominent example of low-end disruption is the emergence of discount retailing, while a prominent example of new-market disruption is the development of personal computers. Notably, the key factor in disruption theory is the ability of disruptive innovations to move up-market, appeal to mainstream customers and, thereby, attack incumbents' shares. This characteristic makes these kinds of seemingly irrelevant innovations interesting to researchers and strategists. It is also the reason why they constitute a significant threat to leading firms in all industries.

Christensen emphasizes that in order to make the correct strategic choices when dealing with innovation-driven change, it is important to be able to determine whether a product is truly disruptive. There are many examples of blindness and biases among incumbent firms that make them ignore or underestimate start-up firms launching disruptive innovations until it is too late.

This is precisely where another justification for digital platforms for data-driven innovation is evident. Many of these new digital platforms have information on start-ups with disruptive ideas. Therefore, the threat discussed by Clayton Christensen is much more important if the innovation modality is closed innovation. However, with the advent of digital platforms covering start-ups, incumbents are now able to detect start-ups with disruptive innovation at a much earlier stage and compare them in segmented cohorts, thereby ensuring that they do not ignore or underestimate them.

Rapid change in an uncertain world

New requirements with regard to the speed of innovation are also reflected in the list of the world's largest companies. 15 years ago, some of the largest and most dominant companies had been market leaders and among the world's largest companies for decades. They included companies like GE, several telecom operators, IBM, and Deutsche Bundespost. At this time, Airbnb, Google, WhatsApp, Facebook, Alibaba, Tencent, LinkedIn, SAP, and numerous other IT-based successes had not yet burst onto the scene.

The average span of time that a company is included on the S&P 500 list has fallen from more than 60 years in 1920 to 17 years today.[7] Another way of looking at these dramatic changes and their potential correlation with innovation is shown in Table 3.2, which lists the 10 companies with the highest market capitalization for the years 2000, 2015, and 2018.

Notably, GE was number one in 2000, but it was only ranked eighth in 2015, and it was not in the top 12 in 2018. Similarly, ExxonMobil was number two in 2000, number four in 2015, and number 11 in 2018. Today, it is far from the top of the list. Nine of the top 12 companies in 2000, including the top four, were traditional, non-tech companies. By 2018, the overall picture had changed to such an extent that the top six companies were IT/tech companies.

Table 3.2 A dramatically different top 12[1]

Top 12 in 2000	Market cap ($B)	Top 12 in 2015	Market cap ($B)		Market cap ($B)
General Electric	474	Apple	710	Apple	896
Exxon Mobil	302	Alphabet/Google	449	Alphabet/Google	782
Pfizer	290	Microsoft	368	Microsoft	682
Citigroup	287	Exxon Mobil	334	Amazon	629
Wal-Ma rt Stores	287	Wells Fargo	297	Tencent	540
Cisco	275	Johnson & Johnson	274	Facebook	521
Microsoft	231	Facebook	272	Berkshire Hathaway	519
AIG	229	General Electric	259	Alibaba	467
Merck	216	JP Morgan Chase	255	Johnson & Johnson	395
Intel	202	Amazon	247	JP Morgan Chase	389
Johnson & Johnson	181	Wal-Mart Stores	230	ExxonMobil	371
Coca-Cola	164	Procter & Gamble	218	Wal-Mart Stores	310

1 Bloomberg valuations and Innosight, 2018 Corporate Longevity Forecast:
Creative Destruction is Accelerating. The table is a rework from Innosight which bases figures on Bloomberg valuations as of 1/18/18.

Clearly the pace of disruptive innovations[8] is accelerating, and this picture could change again in a short span of time. More Chinese companies are expected to take center stage, and we may see Indian companies added to such listings. Interestingly, not a single European company is top ranked. This division between largely traditional companies and high-tech companies may also be representative of two different lines of thinking when it comes to innovation.

Many incumbents fail to effectively deal with innovation. Instead, they think about innovation linearly and in a closed, incremental fashion. Consider, for example, the thinking of Nokia, formerly the global market leader in mobile handsets. For many years, Nokia maintained its innovation culture, which had strong Finnish roots. This made cross-boundary thinking difficult. In their award-winning book, *Ringtone*, Yves Doz and Keeley Wilson describe a "cultural misfit" in Nokia combined with a strategic vacuum at the top, an inefficient resource-draining matrix, and challenging geographical factors.[9] It all amounted to an inefficient innovation culture. Years ago, SAP successfully moved a considerable part of its R&D activities to California. The counterfactual question indirectly posed by Doz and Wilson is whether Nokia could have survived in the mobile phone market if they had done the same.

In addition, Nokia's complacent "We are the best. We are the market leaders" culture reflected its overconfidence.[10] This is evident in the following statement from Nokia's Chief Strategy Officer Anssi Vanjoki:

> The development of mobile phones will follow a similar path to that followed by PCs. . . . Even with the Mac, Apple attracted a lot of attention at first, but they have remained a niche manufacturer. That will be their role in mobile phones as well.[11]

Firms and innovation

Nokia's Chairman of the Board, Risto Siilasmaa, reflected on the challenges the company faced:

> In a flash, I saw the root cause of what had unnerved and confused and worried me for the past two years. A virus has been spreading throughout the culture and the company, shielded by our past success, feeding on fear of failure, rebuffing bad news, and suppressing a sense of accountability.[12]

Interestingly, Nokia had a very large R&D budget. It kept adding to that budget without looking at the base. At the same time, the company underestimated the value of external ideas. There were also other deficiencies including an outdated software platform based on Symbian, and the insufficient innovation and execution of a new platform known as MeeGo. Yet again, it seems as if ideas from the outside were disregarded. For several critical years, Nokia refused to rely on the Android system, which was far more advanced and more widely used. Here, we again encounter the "not invented here" syndrome.

In the Nokia case, innovations believed to be rational and predictive by innovators who were thinking conventionally and overconfidently were based on past successes, while external and potentially disruptive ideas were ignored. This is what we call "linear innovation thinking," which is also often inherent in companies utilizing existing modalities to create and adopt innovations.

In the following chapters, we will take a closer look at the new digital platforms. In particular, we will analyze whether they are just an appendix to closed or open innovation, or a completely new modality.

Key reflections

A   In Chapter 3, we addressed the differences between closed and open innovation, and we uncovered certain advantages and disadvantages. Open innovations offer higher effectiveness but lower efficiency, while closed innovations are associated with higher efficiency and lower effectiveness. Data-driven innovation from the new digital platforms if characterized by high effectiveness and high efficiency. *Can data-driven innovation serve as a substitute for closed and open innovation? Does it constitute a new modality? Alternatively, should data-driven innovation be seen as a toolkit for "fixing the engine" in relation to the deficiencies of closed and open innovation?*

B   A book on innovation must also address the issue of disruptive innovation. *How should data-driven innovation to be viewed in this context? Does it accelerate or decelerate disruptive innovation?*

C   We are living in an era of increasingly rapid change. This affects the rankings of the largest global companies when measured in terms of relative market capitalization over time. The top five in 2000 have been replaced with a set of five companies based on data-driven business models. *Why has data-driven innovation not yet played a significant role in the creation and adoption of innovation?*

Notes

1 Andersson, U., Dasi A., Mudambi, R. and Pedersen, T. (2016). Technology, Innovation and Knowledge: The Importance of Ideas and International Connectivity. *Journal of World Business*, 51(1): 153–162.
2 Von Hippel, E. (1988). *Sources of Innovation*. New York: Oxford University Press.
3 Chesbrough, H. W. (2003). *Open Innovation: The New Imperative for Creating and Profiting from Technology*. Boston, MA: Harvard Business School Press.
4 Laursen, K. and Salter, A. (2006). Open for Innovation: The Role of Openness in Explaining Innovative Performance Among UK Manufacturing Firms. *Strategic Management Journal*, 27(2): 131–150.
5 Foss, N., Laursen, K. and Pedersen, T. (2011). Linking Customer Interaction and Innovation: The Mediating Role of New Organizational Practices. *Organization Science*, 22(4): 980–999.
6 Christensen, C. L. (1997). *The Innovator's Dilemma: When Technology Cause Great Firms to Fail*. Boston, MA: Harvard Business School Press.
7 International Board of Innovation (2019). *Guide of Innovation Transformation*. https://www.board ofinnovation.com/guides/innovation-transformation-guide/.
8 Makides, C. (2008). *Game Changing Strategies: How to Create New Market Space in Established Industries by Breaking the Rules*. San Francisco, CA: Jossey-Bass. Makides coined three different kinds of disruptive innovation: disruptive technological innovations, disruptive product innovations, and disruptive business-model innovations. We would argue that there is a considerable degree of overlap between these three types.
9 Doz, Y. L. and Wilson, K. (2018). *Ringtone: Exploring the Rise and Fall of Nokia in Mobile Phone*. Oxford: Oxford University Press, cf. p. 148 and appendix 1.
10 Ibid.
11 "7 Quotes From Top Corporate Execs Who Laughed Off Disruption When It Hit", www.vergify. com.
12 Siilasmaa, R. (2019). *Transforming Nokia: The Power of Paranoid Optimism to Lead Through Colossal Change*. New York: McGraw-Hill Education, p. 72.

chapter 4

Data-driven innovation might be part of the solution

One practical way to address innovation is to look at start-ups worldwide as many start-ups work with combinations of new ideas, new business models, and disruption. Moreover, many start-ups are front-runners in developing new technologies because they are less biased by the past. Interestingly, a number of platforms have recently emerged that offer access to innovations developed by start-ups. As significant amounts of data are collected on these platforms, we see a new industry developing that revolves around digital platforms for data-driven innovation. Finding the "right" piece of innovation is sometimes like finding a needle in a haystack. Such exercises are not easy, and they are often more emotional than rational. Therefore, it is only natural that a number of digital platforms have emerged in order to untie the Gordian knot of matching the different qualifications of corporations and start-ups. One of many advantages of relying on digital platforms is speed in identifying innovative ideas. While some traditional methods may be cumbersome and time consuming, it is easier and faster to undertake searches using qualified databases.

As we have seen in the previous chapters, there are severe cognitive and emotional constraints when it comes to innovation. Therefore, this chapter addresses how data-driven innovation may help overcome some of these cognitive and emotional constraints.

Build or buy – ways to acquire and access innovation

Corporations have traditionally produced much of their innovation in-house. However, as we saw in the previous chapters, one of the most prominent options when trying to colonize the unknowns and achieve innovation is to acquire innovation from vibrant start-ups.

Nevertheless, while 84.9% of corporate strategy executives say that innovation is very important, 78% of them only focus on incremental changes. This is hardly "innovation." By focusing on incremental change, resources and activities are concentrated on continuous and mediocre closed innovation rather than on open innovation with its disruptive threats and opportunities. Moreover, continuous innovation is then understood as simply enhancing existing products or services, cutting costs, and redirecting efforts toward productivity.[1]

Various surveys on the topic of whether to make or acquire innovation exist but none of them are linked to the criterion of success. Consequently, we will discuss several of these surveys' highlights, as we have already concluded that there is an urgent need to prioritize buy options (i.e., external input) in the future.

Today, most corporations seem to have a rather insular view of how to deal with innovation. When a CBInsights (2018) cohort of 677 corporate strategy executives were asked how they would rank their companies' orientations toward innovation, 51% indicated that they would rather innovate internally (i.e., the insular view) than acquiring innovation (18%). The remainder were in favor of innovation based on partnerships (31%).

As corporations prefer to build rather than acquire, corporate innovation is slow. In the same survey, 60% of companies said that it took at least a year to create new products, with 25% saying it took at least two years from ideation to launch.[2] We conclude that only a small percentage of corporations (less than 20%) even consider the "buy" option. This is probably also the case with other types of organizations such as venture-capital firms and investment banks.

There is considerable overlap between closed innovation on the one hand, with conventional corporations preferring an internal make/build option, and open innovation on the other hand, with a few "out of the box" corporations favoring the buy option.

Arthur D. Little conducted a comprehensive survey of relevant issues in this context.[3] These results indicated that innovation is one of the top three strategic priorities among corporate leaders and that this perceived importance is increasing. Moreover, the survey showed that start-ups are here to stay and that they play an important role in corporate innovation. Such innovation takes place in various forms of collaboration, such as events, incubators, accelerators, "corp-ups" (corporate and start-up collaboration), investments, and acquisitions – primarily initiated by the corporations.

An interesting finding is that companies eventually shift away from equity-based approaches in which they either invest in start-ups or simply acquire them. This benefits "softer" forms of collaboration, a conclusion Arthur D. Little bases on responses to the question of whether ownership is a pain or a gain.

Unfortunately, the study does not address the effects of data-driven developments in this context, especially how the advent of intermediaries and platform suppliers can facilitate this collaboration or serve as a new path that may allow corporations to gain access to innovation. In order to develop this idea further, we look at the importance of the start-up philosophy to corporations, and how data-driven innovation can alter the build or buy equation in favor of the "buy" option, where access to innovation is gained instantly.

The importance of the lack of focus on the possibilities of data-driven innovation can hardly be underestimated. The value of a paired comparison between the accelerator way and the data-driven way can hardly be underestimated.

Data-driven innovation might be the solution

Table 4.1 Comparison of innovation through corporate accelerators and digital platforms

	Corporate accelerator	Digital platforms (AI)
Timeframe	one year or more	A matter of weeks
Innovation input	Typically 25 start-ups	An indefinite number of start-ups
Methodology	Manual, serendipity-based	Data-driven, AI assessments
Selection	Gut driven	AI-based selection
Environment	Physical, event based	Virtual, digital, platform-driven

Table 4.1 provides an interesting comparison of data-driven innovation from digital platforms and innovation through one method of open innovation – corporate accelerator programs. Corporate accelerator programs are characterized by analogue workflows. In contrast, data-driven innovation from digital platforms constitutes a digital workflow and supports other digital workflows.

Conventional and disruptive innovations – simulating start-ups

Practically speaking, all corporations would claim that they invest in innovation. Therefore, a slightly different question needs to be asked – one that centers on the approach to disruption. Most corporations worry about disruptions, but statistics suggest that few invest in preventing them.[4] This basically leaves us with two types of corporations.

Type one corporations (most corporations) develop closed innovations in-house. They do so incrementally. They avoid becoming a first mover and see themselves as fast followers at best. They try to mitigate risks and deal with innovation on an insular, ad-hoc basis with slow, short-term innovation processes. This type of corporation is unlikely to acquire innovation and is happy with the known universe. Consider, for example, the following statement by Christopher Norton, Executive Vice President of Global Product and Operations at Four Seasons, one of Airbnb's competitors: "Our guests don't want the Airbnb feel and scent." Even more alarming is a comment made by Richard Jones, Senior Vice President and Chief Operating Officer of Hospitality Ventures Management Group in a 2014 interview with *Hotel News Now*: "We have not seen a direct effect [from Airbnb] in any of our hotels. . . . We don't feel it's having any impact on our results or that it has hit our radar as of yet."

Such statements stand in sharp contrast to Airbnb's overwhelming subsequent success. Within the span of five years, Airbnb's basic concept set a new bar with four million people spending the night at a listed property. Moreover, the company disrupted the hotel chains by introducing higher-end offerings such as Airbnb Luxe. This concept encompassed innovations on the typical hotel chain offering including fridge stocking, tailored childcare, personal chefs, and local activities in addition to airport transfers. Interestingly, much of what now constitutes Airbnb's success might well have been spotted using digital platforms had the traditional hotel chains not been so complacent within the world of the known-knowns.

Type two (few corporations) are high performers, first movers, and risk takers. They invest in disruptive projects and are characterized by non-bureaucratic organizations. As Steve Jobs put it when he returned to Apple:

> What was wrong with Apple wasn't individual contributors. We had to get rid of about four thousand middle managers. Good technical people stepped up to become managers. The way you grow at Apple is not the same as at GE.[5]

Apple tried to create very small teams and assign them to specific tasks, thereby simulating a start-up company. According to former Apple designer Andrew Borovsky:

> If two to four people can get it done, you don't need twenty to thirty, which is how so many other companies do it. At Apple, really small teams work on really important projects. That is one of the advantages of being a start-up.[6]

Yet again, we find considerable differences between conventional corporations and corporations that either simulate a start-up context or learn from start-ups. Steve Jobs commented on this, stating:

> One of the things that happens in organizations as well as with people is that they settle into ways of looking at the world and become satisfied with things. And the world changes and keeps evolving and new potential arises, but these people who are settled in don't see it. That's what gives start-up companies their greatest advantage.[7]

Generally, type two companies are more inclined to engage in open innovation and to obtain input from external sources, especially start-ups. This leads us to an examination of data-driven innovation.

What is data-driven innovation?

We do not suggest that all innovation can easily be purchased externally – that would essentially be the same as pleading for complete outsourcing of corporations' innovation and R&D. Rather, our aim is to point to the need for a new balance between build and buy, as the buy option is often either overlooked or underprioritized.

Many possibilities exist with regard to the buy option. There are already a considerable number of platform vendors through which corporations can gain access to start-ups and, thereby, innovation.

The exponential development of big data is well known. The curve speaks for itself (see Figure 4.1).

Within the big data space, patterns can be detected in the data. For example, Google was better at predicting the spread of the SARS virus than the World Health Organization, the supposed expert. Google simply assumed that the origination of searches from a geographical perspective was identical to the actual spread of SARS. The correlation was high, although one could question whether there was causality.

Data-driven innovation might be the solution

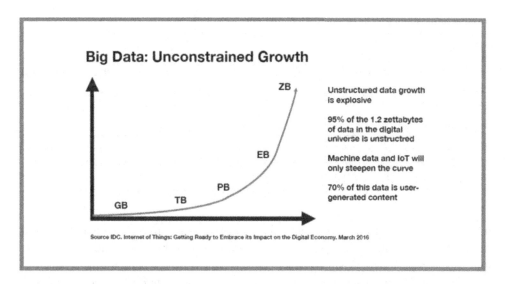

Figure 4.1 The big data explosion

A completely different situation is evident with regard to innovation, where there is no easy method of measuring it such as Google searches, IBM's Watson database or other databases. In fact, for many years, there was no database on innovation.

Today, one can search the internet for "innovation." However, what comes out of such a search is less than adequate. Whether you use search engines or AI, the result is highly dependent on the quality of the data. If you, for instance, google "innovation," you will receive no less than 2,150 million hits. You can reduce this by a factor of 31 if you search for "innovation water pumps," which results in only 68.5 million hits. The number can be furthered dramatically by searching for start-ups within this field, recognizing that much modern innovation comes from start-ups.

Interestingly, a number of platforms have recently emerged that offer access to innovations developed by start-ups. As significant amounts of data are collected on these platforms, we see a new industry developing that revolves around digital platforms for data-driven innovation. One of the major advantages of digital platforms is that they extend the search for start-ups and innovation beyond firms' own collaborators. While open innovation is typically conducted in collaboration with close counterparts like suppliers and customers, data-driven innovation platforms search for innovative start-ups all over the world. As such, they move open innovation to a new level, where the search for innovation becomes global.

Therefore, we need to take a closer look at this type of platform. One underlying rationale for building these platforms lies in the obvious advantages of collaboration between corporations and start-ups, which are captured in Table 4.2.

As illustrated in Table 4.2, complementarity exists between innovation, resources, customers, and scale. Corporations have what start-ups lack, and start-ups have what corporations need. How should this general match be supported?

The answer to this question was blowing in the wind for several years during which experiments were launched with a number of modalities such as start-up events,

Table 4.2 Complementarity between start-ups and corporations[1]

Start-ups	Corporations
Innovation-centric and high growth potential	Lack of growth and innovation
Lack of resources	Ample resources
Lack of customers	Strong customer base
Lack of scale	Large scale

1 This table and some of the tables and text sessions in the following section are based on Andersen, M. M. and Pedersen, T. (2020). *From Innovation to Exnovation: How Data-driven Innovation Platforms Change the Game*, Working paper, Copenhagen Business School and Bocconi University.

incubator and accelerator programs, direct investments, partial investments, and M&A activities. Most of these initiatives were introduced by corporations. However, none of them were data driven.

Finding the "right" piece of innovation is sometimes like finding a needle in a haystack. Such exercises are not easy, and they are often more emotional than rational, especially when the traditional modalities are utilized. Therefore, it is only natural that a number of digital platforms have emerged in order to untie the Gordian knot of matching the different qualifications of corporations and start-ups.

One of many advantages of relying on digital platforms is speed. While some traditional methods may be cumbersome and time consuming, it is easier and faster to undertake searches using qualified databases. Speed here goes hand in hand with effectiveness. Instead of contacting a limited and serendipitously selected number of start-ups located nearby, corporations can search among millions of starts-ups globally and in real time.

Given these advantages and the emergence of new platforms, we now examine the inner workings of some of these platforms.

Digital platforms for data-driven innovation

One practical way to address innovation is to look at start-ups worldwide as many start-ups work with a combination of new ideas, new business models, and disruption. Moreover, many start-ups are front-runners in helping societal transformation take place and in developing new technologies. Often, their work is facilitated by digital rebels, AI, or big data. Numerous platforms with data on start-ups have been established with a regional focus, typically within a specific country. This may have meaning for purposes other than innovation, such as when a country needs to establish a register of start-ups within its borders, or in order to follow up on investment schema, tax incentives, and national support for start-ups. However, as we are focused on the acquisition and appropriation of innovation from start-ups, only truly international, cross-border data make sense. Innovation is not confined to national borders. Ideally, platform vendors should be comprehensive and global in their reach.

Several vendors are looking into this space with the aim of making this important information available. Based on the degree of digitalization and tailor-made offerings,

Data-driven innovation might be the solution

two types of vendors are identifiable. One group provides platforms that rely on the usage of AI and machine learning, while the other group is more focused on platforms that enable the delivery of consultancy services as well as industry and trend analyses.

When we use the term "platform," it is interchangeable with "framework." This means that this chapter addresses frameworks and platforms more than tools, which are the subject of Chapter 5.

AI-based platforms

For a number of reasons, there are considerable differences between platforms that are generally based on AI and those that are not (or those with less advanced AI capabilities).

As Table 4.3 indicates, there is a broad variation in the available AI-based platforms (a more detailed presentation is provided in Appendix C). The vendors can be grouped according to their focus areas. Some are focused on VC investments/funding (e.g., Hatcher, Tracxn, Motherbrain, and Funderbeam), while others are mainly focused on corporate innovation (e.g., Valuer.ai and Innospot). In terms of their data-collection processes, some platforms focus on both qualitative and quantitative techniques, with distinct analyses and content decks offered for each start-up (e.g., Valuer.ai). Others merely look at existing databases in order to mechanically generate output (e.g., Tracxn). Finally, the artificial intelligence of the different platforms focuses on optimizing different objectives ranging from deeply curated innovation (e.g., Valuer.ai) to replication of primarily quantitative data from other databases (e.g., Hatcher, Motherbrain, Preseries, and Tracxn). The different focus areas also represent the different approaches in how corporations deal with innovation processes in a data-driven world.

Model A (Figure 4.2) deals with contingencies in which there are opportunities to take a deep dive into early-stage start-ups using qualitative techniques in addition to the usual quantitative data. This represents a comprehensive platform, which requires a great deal more than just integrating existing quantitative databases. Appendix B provides an elaborate example of the complex flow of a digital platform with advanced AI capabilities.

Table 4.3 Platforms offering information on data-driven innovation

Name of the vendor	Main focus
Valuer.ai	AI platform with proprietary qualitative data collection
Preseries	Tools to evaluate and discover start-ups
Hatcher	A framework for VCs to discover start-ups
Tracxn	Quantitative framework regarding corporations/VCs
"Motherbrain"	Internal EOT framework
Funderbeam	Focused on funding
Innovationscout	Scouting platform for corporations
Findest	Scouting platform utilizing Igor AI
Innospot	AI based on scouting technology

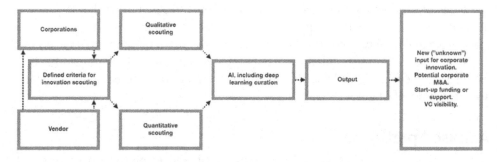

Figure 4.2 Data-driven innovation, Model A

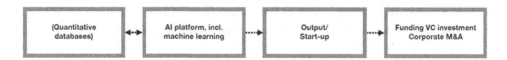

Figure 4.3 Data-driven innovation, Model B

Model B is far less comprehensive when it comes to data collection. It is suited for vendors preferring to rely on existing data or general quantitative surveys. A model of this type of data-driven innovation is presented in Figure 4.3.

In an analysis of more than 50 vendors, Valuer.ai was the only vendor using both human and artificial intelligence for data collection and analysis as of early 2020. This approach creates platform capabilities with a direct focus on innovation and with the possibility of combining each customer's request with available platform data as well as fresh and focused data collection and analysis. The digital workflow also enables the customer's teams to "train" the data (see Appendix B). A number of other vendors have interesting AI platform capabilities but with a focus other than delivering data-driven innovation to corporations. This may also reflect the fact that there is a lucrative market for other services that revolve around, for instance, M&A transactions.

Valuer.ai's platform uses unsupervised learning to construct a multi-dimensional space in which all start-ups are clustered. Start-ups with similarities are closer together in this multi-dimensional space, which is the foundation of the platform's objective and data-driven selection model. Natural language processing is then applied to the client's request, which is converted into a multi-dimensional vector. The vector is projected into the multi-dimensional cluster, which identifies start-ups in close proximity to the client's focus. The platform automatically identifies and selects relevant start-ups and presents the preliminary results to the client, who then gives feedback on the suggestions through the platform. The client-feedback mechanism is part of Valuer.ai's practically applied reinforcement learning, which triggers adjustments in the client's search. This process is repeated until satisfactory fit is achieved.

Data-driven innovation might be the solution

Platforms and services without major AI capabilities

There are also a number of vendors with solutions based primarily on human intelligence rather than AI. A number of these vendors are listed in Table 4.4.

Table 4.4 illustrates the presence of a considerable variation in platforms and services without major AI capabilities. They can be subdivided into distinct subgroups. One group is focused on delivering, for instance, start-up competitions and acceleration programs (e.g., youNoodle and Accelerace). Another group revolves around funding and investment issues (e.g., Lets Venture, Pitchbook, and Seedinvest). Some are focused on niche specialties within the wider environment of funding and investing. Their activities include the rating of start-ups (e.g., OddUp), trading of pre-IPO employee shares (e.g., EquityZen), financial intelligence (e.g., Privco), and enabling individuals to access angel investing as an asset class (e.g., GoBeyondInvest). Finally, some platforms are cross-boundary (e.g., CBInsights) and/or focused on data-driven innovation in general (e.g., Avanto Ventures and Index.co).

Numerous vendors in this category offer, for example, access to smaller databases or consulting services. A few of the companies listed are neither particularly data-driven nor digital at this stage (e.g., Avanto, Accelerace, and WhataVenture).[8] However,

Table 4.4 Platforms and relevant services without major AI capabilities

Name of the vendor	Main focus
CBInsights	Tech-market intelligence platform based on masses of data and a variety of data sources
YouNoodle	Connects corporations, start-ups, and entrepreneurs through online competitions and challenges
Accelerace	Accelerator programs for start-ups and corporations
Avanto Ventures	Help Nordic customers venture into new growth areas by matching them with the start-up ecosystem
Privco	Private company intelligence
Equity Zen	An online marketplace for trading pre-IPO employee shares from privately held companies
OddUp	A platform to generate start-up ratings on valuation, etc.
GoBeyondInvest	Platform enabling individuals to access angel investing as an asset class
What a Venture	Innovation consultancy and expert in guiding new business ideas to success
Index.co	Enriches and automates workflows regarding data on high-tech companies
Seedinvest	Equity crowdfunding platform
LetsVenture	Online funding platform that enables start-ups looking to raise seed capital to create investment-ready profiles online, and connect to accredited investors
Pitchbook	A web-based data software, research, and analysis tool that provides information about VC, PE, and M&A transactions, including public and private companies, investors, funds, firms, and people

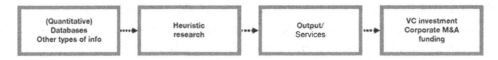

Figure 4.4 Data-driven innovation, Model C

when the focus is directly on accessing data-driven innovation, one company stands out owing to its broad, cross-boundary platform: CBInsights.

CBInsights is a macro-level platform that provides data on market specifics in the technology sector. The data and analyses it provides give customers an overview of market specifics such as competitor strategies, market trends, market size, transactions, and company investments as well as the sourcing and valuation of deals. The company aggregates massive amounts of data, undertakes research, and organizes this data, often using visualization tools.

Crunchbase[9] describes CBInsights as follows:

> CB insights is the ideal tool for those engaged in private equity, venture capital, corporate development, investment banking, corporate innovation & strategy, angel investment and consulting. Whether it is deal sourcing, due diligence, or market and competitive intelligence, CBI has been created to assist you in discovering the right private company information in the most efficient, comprehensive way.

The promotional wording is well deserved when it comes to the broadness of the platform offerings, which are built on the basis of an organization in which a series A round of financing has already transpired.

In continuation of the innovation models described in Figures 4.2 and 4.3, we now address a third model, which is the model largely adopted by CBInsights. This model is based on massive amounts of data, heuristic methodologies, and output for industrial buyers, like VCs and actors involved in corporate M&A activities. Model C (Figure 4.4) deals with innovation in much the same way as CBInsights.

In this model, massive amounts of data are collected, such as data from other databases and new data created by general surveys. The data are utilized at the platform level for heuristic research. The output in the form of various subscription services is also relevant for specific investment purposes, with VC investments and corporate M&A seemingly having the highest priority.

How do CBInsight and Valuer.ai compare?

In light of our examination of the different vendor types, their platforms for delivering services, and our underlying understanding of how to generate innovation, two vendors seem particularly interesting when it comes to the notion of data-driven innovation: CBInsights and Valuer.ai. We will therefore look into a paired comparison of these two vendors.[10]

Data-driven innovation might be the solution

Customer examples

The first focal point is customer focus. The two companies' claims suggest they are similar in this regard – both companies mention start-ups, corporations, and venture capital. However, a closer look reveals that Valuer.ai is focused on bringing corporations innovation from earlier stage start-ups using well-defined search criteria from the corporate client. Conversely, CBInsights' prime client base is VC companies. It aims to make capital markets, especially VC companies, aware of innovation from late-stage start-ups. To some extent, these differences are reflected in the two company's customer lists (based on their websites at the time of writing).

As Table 4.5 indicates, Valuer.ai does not include any VC companies among its key clients. By contrast, more than half of the customers on CBInsights' list are active in the VC environment. This client structure reinforces the impression that Valuer.ai is focused on delivering data-driven innovation services to the corporate segment, while CBInsights is focused on delivering data and analyses on important investment opportunities in innovative, growth-potent companies to the VC environment. This does not prevent the two companies from providing a variety of other services. However, are their different focal areas related to differences in their underlying platforms?

How do the platforms work?

After a client request is accepted, Valuer.ai's platform disseminates the request using relevant apps to its network of around 5,000 international scouts, who are tasked with discovering and detecting innovation in new and relevant start-ups. At the same time, AI-based searches are launched on start-up innovations already included in Valuer.ai's database. The clustered start-ups are exposed to client-specific searches, called "radar calibrations," which process the search definition using a neural network in order to match all relevant candidates. Valuer.ai applies natural language processing to the free-form text description provided by the client and converts all provided information

Table 4.5 A selection of customers

CBInsights	Valuer.ai
Sequoia	ABN Amro
Castrol Innoventures	Radiometer
NEA	BMW
500	Grundfos
Upfront Ventures	Spirent
Homebrew	Novo Nordisk
Redhat	Siemens
Telefonica	Novozymes
KPMG	Uniper
Cisco	Pfizer

Figure 4.5 An illustration of platform-based data-driven innovation

into a multi-dimensional vector. The vector is projected into the multi-dimensional cluster, which then identifies start-ups in close proximity to the client's focus. In practical terms, this search projection acts as an "ideal" start-up for the client and is then compared to possible candidates.

New start-ups are added to the platform by way of a company-specific content deck, which is based on a combination of quantitative and qualitative data. Most of the quantitative data comes from existing databases, while the qualitative data are usually generated through company-specific interviews. These interviews are repeated over time in order to track developments and to allow for the use of time-series analyses. Deep learning within the AI platform allows for the assessment of qualitative factors such as talent, management capabilities, competitive strengths, and weaknesses. Most of this data is generated by the AI platform itself, while the qualitative part of the first interview and the establishment of that data within the content deck is handled by analysts and assessed by researchers. A global team of experts provides a second opinion. The entire process is illustrated in Figure 4.5.

The six steps may be illustrated as follows:

1   The client completes the Radar request (search request) by defining the aim and overarching goal through the interactive platform. The platform calibrates and prepares the Radar request for start-up clustering and activation.
2   The calibrated Radar (search definition) activates the agent network (the crowdsourcing community) and the database to continuously identify and cluster candidates. The resulting matches (relevant start-ups) within the clusters serve as the foundation of the supervised learning algorithm applied in the next stage of the process.
3   The platform empowers client stakeholders to democratically vote and evaluates the identified candidates to calibrate an optimal focus. These activities trigger and adjust the Radar. Depending on the outcome of the first democratic vote, there may be several iterations until a satisfactory Radar calibration is achieved.
4   The extensive data-enrichment process qualifies the candidate's data through dynamic scoring. The candidates with the highest scores from the AI scoring and matching are evaluated through research and interviews in order to further enrich the data and determine, for example, the 25 most relevant candidates.
5   In the case of Valuer.ai's platform, they engage industry expert networks to validate the matched candidates. The validation provides a holistic overview and ensures an objective perspective on the matches. Validation is the final input for the ranking and adjustment of the candidates identified through the Radar.

Data-driven innovation might be the solution

6   The matches are made available on the Valuer.ai platform along with, for example, 10 highlighted candidates. The information is provided in a digital insight report based on expert insights and client evaluations. The data are continuously renewed to ensure an up-to-date overview. The platform then facilitates evaluation and recommends recalibrations that enable the continuity of the Radar.

Valuer.ai's bottom-up model serves as the foundation for a new macro approach to supporting the different needs of its clients. The searches undertaken on behalf of Valuer.ai's clients have created a large pool of structured data that is used to identify trends in relevant information. The Valuer.ai database is clustered as part of a macro analysis in order to identify innovation trends, and cross sections of relevant sectors are mapped onto subcategory trends and drivers. The macro-database analysis concludes with a foundational prediction for the most relevant radar calibrations. This is an example of how the combination of structured data and machine learning can create additional features that address a larger proportion of clients' needs within innovation.

While Valuer.ai's platform is based on micro-level information and on a built-up, bottom-up principle, CBInsights' platform is a macro-level platform with massive amounts of data based on aggregation. It utilizes analysts and researchers, and provides reports with a very high level of visualization. In brief, the "market intelligence platform analyzes millions of data points on venture capital, start-ups, patents, partnerships and news mentions to help you see tomorrow's opportunities, today."[11]

Essentially, CBInsights is more aligned to partnerships. It offers an impressive amount of market and industry research as well as information on specific industries. As its platform is not designed to work from the "bottom up" like an AI platform (e.g., Valuer.ai's platform), it is interesting to examine its specific features (see Table 4.6).

Table 4.6 An overview of CBInsights' features

Private and public company data	Reports on everything that is publicly available
Mosaic scores	Analyzing health and growth of private companies based on momentum, market, and money
Industry analytics	Sense industry movements, revenues, and trends
Core intelligence	News coverage on private/public companies – scraping data from news sites
Stories	Data visualization with access to edit
Tailored reports/year	Tailored quarterly reports on the market
In-person brief	In-person brief from executive per annum
Events	Discounted tickets/access
Notes	Leave opinions and analysis on any page visible to entire org.
Intelligence unit	Map of opportunities and risks of emerging technologies
Market map maker	Provides "real-time" overview of competitive landscape. www.cbinsights.com/market-map-maker
Transcripts	Public company earning transcripts. Overview of top exec meetings on earning and potential growth sectors

In summary, Valuer.ai's micro-level "discovery" focus goes well with the content decks for curated start-ups on its platform. The innovation platform contains a great deal of "thick" data, which, in combination with the advanced utilization of AI, allows for winning innovations to be uncovered.

In contrast, CBInsights deals more with macro-level "thin" data, which allows for analyses of relevant market and industry trends. Consequently, CBInsights is naturally designed to primarily deal with later-stage companies with a client base. VC companies are its key target. For example, CBInsights has been particularly instrumental in describing, predicting, and following the "unicorn" universe. Probably the most advanced feature is a score for the health and growth of private companies (see the textbox from CBInsights).

Mosaic is a quantitative framework to measure the overall health and growth potential of private companies using non-traditional signals.

The Mosaic score is comprised of 3 individual models – what we call the 3 M's, each relying on different signals (although all the signals utilized are not revealed for obvious reasons).

Market

The quality of the market or industry a company competes in is critical. If you are part of a hot industry, that serves as a tailwind to push you along. Conversely, being in an out of favor space means fewer investors, partners, media, and more.

The market model looks at the number of companies in an industry, the financing and exit momentum in the space, and the overall quality and quantity of investors participating in that industry.

Money

The money model assesses the financial health of a company, i.e. is it going to run out of money? Our model looks at burn rate, the quality of the investors and syndicate that may be part of the company, its financing position relative to industry peers and competitors, and more.

Momentum

The final model is momentum, where we look at a variety of volume and frequency signals including social media, news/media, sentiment, and partnership and customer momentum.

We look at these on an absolute and relative basis versus peers/industry comparables. The relative piece is critical as it ensures that, for example, enterprise software companies. "who may get less media attention or who spend less time on social media are not penalized versus consumer-focused tech companies."

Can you pick the winners or the winning innovation?

CBInsights' system for analyzing the health and growth of companies is called the "Mosaic score." It is based on the measurement of (innovative) companies against three variables: market, money, and momentum.

Interestingly, CBInsights has proven that it understands unicorns (i.e., companies that will eventually be valued at USD 1 billion or more). Given its advisory and research focus related to investments, its scoring tool is particularly interesting for VC companies, investment banks, and similar organizations. In 2015, CBInsights worked with the *New York Times* to predict 50 future unicorns. As of June 2020, 24 of them have hit that mark (48%).[12]

Through its database, which stores data on more than 30 parameters like funding cycles and employees, Valuer.ai has proven that it can predict the likelihood of start-up success. It does so by utilizing its extensive database of historic data to train its neural networks to predict the most likely outcome for a start-up (success or failure). This enables the platform to predict and filter start-ups in terms of their likelihood of success in their industry and to point out start-ups with a fair chance of success to their clients.

One of the main ideas behind any of these AI platforms is to find winners and avoid losers. In this regard, Valuer.ai benefits from having studied, analyzed, and curated start-ups both quantitatively (which a number of other vendors do) and qualitatively (which no other vendor does on the same systematic basis). This helps increase the training and assessment of the AI through supervised, deep-learning algorithms based on training data. Calculations based on historic data from Valuer's platform predict the likelihood of any start-up success with 85% certainty. This is also why Valuer.ai has built such elements as crowdsourcing networks and start-up interviews into the model. These elements enable it to achieve the data quality required to fully utilize the prediction models.

Pricing the services

A like-for-like comparison of the prices is difficult for several reasons. First, the underlying products and services differ substantially. Second, prices may not be stable over time. Third, Valuer.ai's prices are not publicly available, while some of CBInsights' general prices are.

Nevertheless, it is still interesting to look at the price differentials (see Table 4.7 and Appendix C).

CBInsights' prices range from US$59,690 to US$265,000 on a yearly basis, whereas Valuer.ai's prices run from US$120,000 yearly for the standard package to US$288,000 for the most comprehensive.

A closer look at the two offerings highlights a considerable difference. CBInsights' offerings seem to be based on a "more of the same" philosophy. Starting with analytics, the client receives offerings that ultimately culminate with trend analyses, metrics, and toolkits likely to be relevant for VC purposes. CBInsights has also recently included a scouting offering in its most expensive offering.

Table 4.7 Price comparison of CBInsights and Valuer.ai (as of May 20, 2020)

CBInsights	Valuer.ai
Annual rates (USD)	Monthly rates (USD)
Executive: 265,000	Standard subscription: 12,000
Insider: 199,400	Premium: 24,000
	Executive: Bespoke/White Label solution
Prime: 99,700	
Answers: 79,900	
Analytics: 59,690	

Valuer.ai's offerings follow another principle. The starting point is for clients (usually corporations) to be offered search missions among highly curated start-ups ("finding a needle in a haystack" innovation) through a standard subscription. In addition, a number of digital advisory packages centered on instructive videos and advisory assistance are available with the aim of helping corporations appropriate as much value as possible out of innovations and to assist them in blitzscaling the value of such innovation.

Moreover, Valuer.ai has commenced making its platform available for wholesale. This means that third parties will be able to access the platform and resell valuable innovations in their own names and for their own purposes. For example, banks can purchase a white-label solution and then resell the information on innovation from Valuer.ai's platform to their own customers in their own name.

Concluding perspectives

The description and analysis of the various platforms show that a new trend is quickly emerging – it is now possible to access innovation from a variety of digital platforms. Regardless of whether we are talking about corporations, VC funds, investment banks, or other actors, the need to deal with innovation is continually increasing in today's uncertain world. Moreover, the number of black swan events and black swan organizations is rising steadily.

The AI models show an interesting and valuable result of the AI innovation process – the inclusive feedback process and the multiple search cycles result in a continuous learning cycle, which improves the platforms' ability to correctly identify relevant candidates for each individual client. This is an example of AI working with humans to capture and apply learning, thereby achieving ongoing and unbiased learning.

Non-AI platforms exhibit a better fit with consultancy and advisory work based on macro-level analyses, analytics, and aggregated data. It is impossible to say whether one type of platform is better or worse than the other. However, a deep dive highlights

Data-driven innovation might be the solution

notable differences in the offerings. Therefore, we deliberately depicted one leading vendor from each of the two groups: Valuer.ai and CBInsights.

Valuer.ai offers Innovation as a Service (IaaS), which is comparable to the business models of cloud-based software solutions sold as SaaS. This means that corporations and others can approach this vendor with a request list and acquire curated innovation. Moreover, matchmaking can be orchestrated between start-ups on the one hand and corporations or venture capital firms on the other hand.

CBInsights does not curate innovation from a large cohort of early start-ups from a micro-perspective to the same extent. Rather, it utilizes all intelligence gathered on its platform to deliver reports at a macro level. Furthermore, enhanced data can be delivered as analytics (e.g., on patents and market size) or as "tailored intelligence," including in-person executive briefings and tailored reports.

Some key characteristics of the two platforms are compiled in Table 4.8.

As illustrated in Table 4.8, CBInsights' and Valuer's platforms are quite different, and they are probably more complementary than competitive in nature.

In the next Chapter 5, we will look into how these offerings are utilized in real life. We will also examine how a crowdsourced view on this issue helped uncover strengths and weaknesses in this context.

Table 4.8 Characteristics of CBInsights and Valuer.ai

Topic	Valuer.ai	CBInsights
Main focus	Start-ups	VCs
Main level	Micro	Meso or macro
Stage of innovation	Early	Later (post-beachhead)
Platform	AI-based Innovation	"Emerging Technology Insight Platform"
AI	Yes, advanced	Beginning to apply machine learning
Two-sided-platform	Yes	No (generally sources data from many vendors, sometimes at an aggregate level)
Data-sourcing method	Crowdsourcing, agents, and qualitative interviews	Existing database on a primarily quantitative basis and subsequently analysts
Main analytical method	AI	Analysts
Qualification of data	Qualitative interviews with all start-ups for curation purposes; additional expert input; solid ground data/ thick data	Basic data with qualitative techniques, but ground data subsequently analyzed by analysts
Subjectivity	Client defines subjectivity through requests; data is handled without personal subjectivity or potential bias	Massive amounts of data are handled through the lenses of analysts who can tailor results to clients and provide specific, personalized advice

Key reflections

A   In Chapter 4, we began with some reflections on build or buy considerations. In our descriptions in Chapter 3, we assumed that innovation was built and not bought in both the closed and open innovation modalities. However, a new trend emerged several years ago in the area of collaboration between start-ups and corporations. Such collaborations can take many forms – they can serve as accelerators or incubators or provide support for M&A activities. A more institutionalized version comes with the digital platforms, several of which are based on start-ups. *What are the pros and cons of utilizing collaborations or digital platforms? Are they mutually exclusive or complementary?*

B   The descriptions of the various platforms indicate that data-driven innovation from new platforms is one way to facilitate the creation and adoption of innovation. *Why is a higher degree of effectiveness and efficiency possible when utilizing digital platforms? Is there an optimal combination of open innovation, closed innovation, and purchases from one or more platform vendors?*

C   The new platforms for data-driven innovation do not provide a homogenous picture. In fact, the key characteristic is the presence of more dissimilarities than similarities. The two vendors we examined exhibit a high degree of complementarity. *Under which contingencies is it preferable to buy data-driven innovation from more than one platform? If one is forced to buy innovation from just one platform, when is it preferable to buy from a platform delivering thick data at the micro level and when is it preferable to buy from a platform delivering thin data at a macro level?*

Notes

1   CBInsights (2018). *State of Innovation: Survey of 677 Corporate Strategy Executives*. Originally from www.cbinsights.com, but see also *Why We Can't Innovate* at https://www.forbes.com/sites/billfischer/2018/05/06/why-we-cant-innovate/#b360a33a55cf.
2   Ibid., p. 25f.
3   Arthur D Little and Match-Maker Ventures (2019). *The Age of Collaboration II: Start-ups + Corporates = Pain or Gain?* https://www.adlittle.com/en/age-collaboration-ii.
4   Ibid.
5   Lashinsky, A. (2012). *Inside Apple: The Secrets Behind the Past and Future Success of Steve Jobs's Iconic Brand*. London: John Murray, p. 71.
6   Ibid., p. 81.
7   Ibid., p. 170.
8   It is fair to discuss which companies should be in and which should be out. We attempted to calibrate this during the interviews, and through our personal experiences with accessing and utilizing some of the platforms. For Michael Andersen, the latter includes participation in one of the underlying companies. Interesting companies with considerable strength in innovation include such companies as Widerpool and Itonics. There are not included here simply because they deliver more holistic services, including innovation management in an organizational context. This is beyond the scope of this book, which concentrates on data-driven innovation from digital platforms.
9   Cf. www.Crunchbase.com (February 12, 2019).
10 Generally, the following is based on what is available on the websites of the two companies as of February 2019. In order to be as accurate as possible, we conducted a few supplementary interviews. Nevertheless, comparing companies with business interests is always a matter of finding a balance. We have tried to strike and test this balance among interviewees.
11 CBInsights website (February 2019).
12 Ibid.

chapter 5

Utilizing platforms with data-driven innovation

Two different types of platforms exist that support data-driven innovation. One is a broad platform, such as the platform offered by CBInsights, which delivers industry reports at the macro level, and provides a number of services relevant for keeping track of the unicorn market, and for the M&A activities of VCs and corporations. The other type of platform, such as the one offered by Valuer.ai, operates solely at the micro level. It focuses on curating a large number of start-ups at an earlier stage, and on detecting innovation and providing information on that innovation to corporations. The advantages offered by these platforms in terms of innovations are: 1) the use of the right search mechanisms and analytical tools means faster and broader search for innovations; 2) a detailed and highly qualified portrayal of innovative ideas in start-ups owing to the big data environment; and 3) access to the platforms are typically priced such that a customer's total cost of ownership is favorable.

Now that we have looked at some of the platforms available today, it is time to analyze how corporations and others can appropriate value from these platforms. A starting point is to take a high-level look at how corporations deal with the notion of innovation. Figure 5.1 illustrates four salient findings.

From CBInsights, we know that innovation is often closed and insular. This implies that the incubation time for innovation is long. From Valuer.ai, we have learned that severe blind spots exist with regard to innovation. For instance, 95% of decision makers are unaware of the option to acquire innovation. However, when presented with the opportunity to do so, 9 out of 10 would like to take advantage of data-driven innovation.

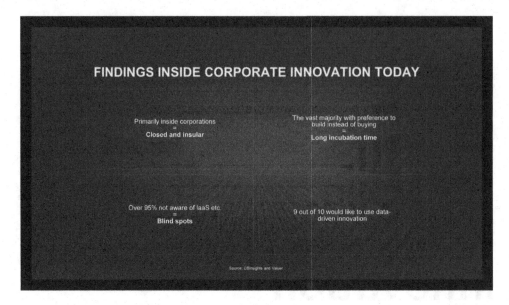

FINDINGS INSIDE CORPORATE INNOVATION TODAY

Primarily inside corporations
=
Closed and insular

The vast majority with preference to
build instead of buying
=
Long incubation time

Over 95% not aware of IaaS etc.
=
Blind spots

9 out of 10 would like to use data-
driven innovation

Source: CBInsights and Valuer

Figure 5.1 Findings regarding corporate innovation

The aggregate of these four findings is reflected in two paradoxes:

- Innovation is primarily dealt with in a closed and insular fashion. However, corporations would like speed and efficiency at the same time, which is not generally achievable. The preference to build rather than buy means that the incubation time is (too) long.
- More than 95% of decision makers are unaware that they can purchase access to IaaS (i.e., purchase innovation as an instant, data-driven service), although 9 out of 10 would like to use data-driven innovation.

Some guidance on tapping into the new world of data-driven innovation seems necessary.

From unknowns to frameworks to tools

As we saw in Chapter 2, innovation deals with the unknown world, regardless of whether we are talking about "finding a needle in a haystack" innovation or the more complex "put a man on the moon" innovation. For corporations, capturing some innovation from outside the organization will increasingly become a matter of survival and a prerequisite for growth. An important part of such innovation can be accessed through or acquired from start-ups.

We also indicated that there are vendors that can supply access to innovation through a number of platforms. Of these, two different types of platforms were addressed. One was a broad platform, such as the platform offered by CBInsights, which delivers industry reports at the macro level, and provides a number of services

Platforms with data-driven innovation

relevant for keeping track of the unicorn market, and for the M&A activities of VCs and corporations.[1] The other type of platform, such as the one offered by Valuer.ai, operates solely at the micro level. It focuses on curating a large number of start-ups at an earlier stage, and on detecting innovation and providing information on that innovation to corporations through IaaS. Over time, the position of these two markets players seems to be converging slightly. CBInsights has introduced a start-up scouting function, but the information provided through this function is not curated or AI-platform-based with deep learning, which is the case for Valuer.ai. Conversely, Valuer.ai has opened up its platform for use in investment decisions by VCs and others. However, it does not offer the same unicorn focus as CBInsights.

Therefore, we will focus on the dissimilarities rather than similarities. After all, our central theme is that new possibilities are emerging to access data-driven innovation with the advent of digital platform players and how this enhances customers' abilities. These new abilities come with a number of advantages, such as:

- *Speedier.* The availability of platforms comprising big data environments with the right search mechanisms and analytical tools means that innovations can be accessed shortly after a search mission has been established. Looking at innovation in a data-driven fashion helps to find "unknown" innovation by mitigating human bias in the process. Although uncovering innovation might require more than one mission, a search mission in a digital platform environment takes a matter of minutes or hours, while "behind the desk" innovation can take several months or years. Moreover, with the latter, you are still working within the known universe and in an analogue fashion.
- *Cheaper.* Access to platforms are typically priced such that a customer's total cost of ownership is favourable. This also relates to the nature of some of the platforms as a kind of collective good with network effects on both the supply and demand sides. Given the economies of scale and scope, the cost discrepancy between building extra R&D capacities into the corporate organization and subscribing to one or more of the digital platforms is significant.
- *Better.* The fact that something can be acquired cheaply and quickly is not enough – it also has to fulfil your purpose and be effective. This is yet another advantage of platforms as they provide a far more detailed and qualified response owing to the big data environment. Moreover, users benefit from the platform's collective learning (e.g., training algorithms and vast datasets). Therefore, if you are looking, for example, for a specific climate technology, searches undertaken through a platform may initially give you 100, then 10, and subsequently the 3 best solutions available globally. Alternatively, if you are looking for a unicorn investment, a number of qualified prospects may come up, thereby facilitating an investment decision.

In this chapter, we will eventually discuss tools and cases for data-driven innovation. In order to arrive at this end, we will address the following question: How do you find the exact type of innovation in which you are interested?

Identifying relevant innovation (the target)

Given the paths of the two platforms we have discussed, one question is whether the best starting point is to utilize the platforms and services along the lines of those provided by Valuer.ai or those provided by CBInsights. There are many potential rationales for subscribing to Valuer.ai's platform. The four rationales that follow take a basic subscription as the starting point.

There is the "finding a needle in a haystack" rationale. This is typically the case when a corporation is looking for innovation in a specific field in which its own R&D activities are insufficient. Examples could include Grundfos wishing to digitalize the water industry globally or Siemens Gamesa desiring to improve its wind-power technologies.

The "instantly having innovation on the radar" rationale is somewhat different. Some corporations just want to ensure they are instantly aware of new developments around the world. For example, some corporations wish to be fully aligned with the newest developments in their industries or in horizontal technologies such as generic AI and blockchain technologies. Others keep tapping maturing start-ups working with technologies that are currently not profitable to adapt but are likely to be of high strategic importance in the future. For instance, start-ups working with such technologies as quantum computing and fusion energy are currently in the very early stages and still focused on research. However, when they mature, they could drastically change major sectors.

The "finding SDG targets" rationale reflects a recent development in which an increasing number of corporations are paying attention to the UN's Sustainable Development Goals. Interestingly, a survey conducted by PwC[2] showed that 72% of companies mention the SDGs in their annual reports, but only 23% disclose meaningful key performance indicators and targets related to those SDGs. One might fear that the actual work on fulfilling the SDGs is less than companies indicate in annual reports. Given this context and the growing international attention on SDGs, it is not surprising that companies like Valuer.ai supply SDG services to allow corporations to combine their core businesses with sustainable business models that benefit more than their revenue-based bottom lines.

For instance, the Valuer.ai platform can be directed at challenges or opportunities associated with specific SDG goals. In such cases, the starting point is the definition of the focal SDG goal(s). The SDGs have definitions that describe the areas that each of the 17 goals represent. These are processed using natural language processing (NLP) to produce a feature hashed version of the goal.

The processed definition of a chosen SDG can then be applied to a cluster of start-ups, which have also been processed using NLP. This standardizes the goal definition and the descriptions of the start-up solutions, which enables a data-driven comparison between the two. Moreover, multiple definitions (sub-goals) of an SDG can be processed and used for a group of start-ups. Each of the sub-goals can be identified in relation to relevant start-ups. The grouped start-ups, with their identified similarities, can then serve as a foundation for identifying start-ups and technologies that are ideal candidates for addressing problems relating to specific SDG goals.

A final rationale is "special requirements for investments in start-ups." This can take a variety of forms. Some investors, such as pension funds, may have stringent investment criteria that are difficult, if not impossible, to meet in the stock market. Such requirements may include a focus on the environment and sustainability, no child labor, fair pay for employees, or a focus on avoiding pollution. Such requirements can be part of a search that will still lead to many good investment opportunities. Similarly, VCs and financial institutions can define specific investment criteria on the platforms.

Most of these rationales are related to large corporations and multinational, if not global, companies seeking to extract value from Valuer.ai's platform, which is based on curated start-ups. However, M&A activities and investments have also emerged as reasons to utilize this platform.

A number of rationales are also evident with regard to CBInsights. The desire to answer the question of "What are the trends in my industry?" is an obvious rationale that CBInsights can accommodate either through its instant industry analyses or on the basis of specific targets. For example, banking institutions wishing to acquire general knowledge on present and future trends can derive value from the various reports on the fintech industry.

The desire to understand "innovative trends across industries" is a more general rationale that correlates well with CBInsights' offerings. The company states: "We identify signals about emerging and disruptive technology and business trends by untangling a mess of unstructured news articles, patents, start-up websites, venture capital financing, and more." Whereas CBInsights does not directly focus on innovation as such, "emerging and disruptive technology" certainly encapsulates a considerable chunk of it.

"I would like to make a cool venture investment" is also among the rationales for acquiring services from CBInsights. As described earlier, this is a specific target group for CBInsights' services, which is also evident in the client testimonials included on the company's website. CBInsights has special competencies when it comes to identifying and tracking unicorns.

Valuer.ai and CBInsights are not the only vendors in this space, as we discussed in Chapter 4. Instead of listing the other active vendors here, we shall look into some use cases.

Use cases

Numerous use cases with considerable variation can be envisaged. Table 5.1 seeks to categorize some of them.

We address each of these use cases in turn.

Table 5.1 Categorizing different types of use cases

	Simple cases	Complex cases
Focus on early stage/micro	A: "Finding a needle in a haystack"	C: "Man on the moon" innovation
Focus on later stage/macro	B: "I would like to make a cool unicorn-like venture investment"	D: "Innovation accelerated through consolidation"

Use case A: "finding a needle in a haystack"

This use case is evident in how a company like Grundfos works to acquire data-driven innovation. Figure 5.2, which is a condensed version of the process, provides a useful illustration.

The process for "finding a needle in a haystack" has a number of phases, as is the case with the platform toolkit from Valuer.ai. In Grundfos's case, one mission could be to identify a variety of start-ups focused on water and energy technologies. Such a mission ("phase 1: radar request") would require the initial alignment of pre-specified mission criteria between the vendor (i.e., Valuer.ai) and the client (i.e., Grundfos), such as innovation objectives, geographical locations, life-cycle stage, and funding challenges. When this step is completed, the next phase (2) is to start the data mining and the search among the 5,000 agents, followed by the collection of candidates for data-driven selection and processing. This allows for the use of scoring techniques and alignment regarding the right types of candidates, which are evaluated using algorithms processing an extensive set of data points. The client's second key involvement in the process occurs at this point, as the client provides feedback that helps train the selection process through "reinforcement learning" (phase 3). Thereafter, the curation and data enrichment commences with the aim of improving the quality of the data on the few remaining candidates and preparing for the final validation and selection (phase 4). A separate phase (5) comprises the industry experts' validation of the top candidates for data-enrichment purposes. In the last phase (6), the top candidates are presented to the client.

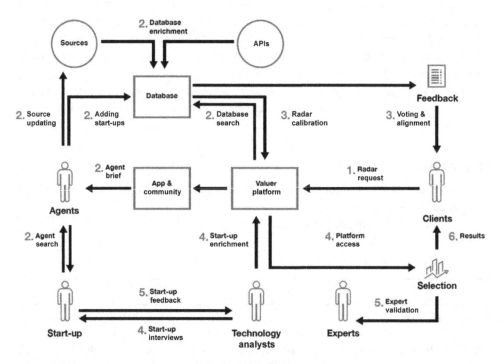

Figure 5.2 Overview of the platform toolkit for data-driven innovation (an extended version of Figure 4.5)

Platforms with data-driven innovation

Grundfos has used the Valuer.ai platform not only for industry-specific purposes with regard to water and energy-related innovation but also in relation to cross-boundary technologies such as AI and blockchain. According to Henrik Juul Nielsen, Senior Manager of Technology and Innovation at Grundfos, "Valuer has added valuable insights to our innovation scouting activities, especially on the edge in radical innovation with AI and Blockchain start-ups" (company website, 2020).

A more detailed presentation of the dataflows is available in Appendix B.

Use case B: "I would like to make a cool unicorn-like venture investment"

In this use case, we envision one of the large venture capital companies (e.g., Sequoia, Norwest Venture Partners, or FirstMark) placing a request for information on potential investment opportunities in a (potential) unicorn or for an assessment of the attractiveness of, for example, a series C or D round. A higher-end subscription to CBInsights would allow for identification of potential unicorns, and the monitoring and valuation of those unicorns, as well as access to information on benchmarks, patenting, and the Mosaic score.

Other platforms could also be approached, such as Valuer.ai or Pitchbook, but it is probably fair to say that CBInsights has a specific competence regarding unicorns. The Mosaic score – with scores related to the market ("Is the target part of a hot industry"?), money ("Will the target run out of money"?), and momentum (a variety of volume and frequency signals, including social media, news media, sentiment, partnerships, and customer momentum) – is key.

Some customers, such as FirstMark, like to discover and track emerging companies, see funding events in real time, and gain insights into nascent industry trends and markets. "We were looking for comprehensive data to ensure we could break down and analyze the segments that matter to FirstMark. We needed the ability to quickly query and interpret the data. CBInsights was the right choice as a result," says resident data expert and the Vice President of Platform at FirstMark, Dan Kozikowski.

Use case C: "man on the moon" innovation

By "man on the moon innovation," we refer to all the innovation that was necessary to put a man on the moon in 1969. The term also highlights the fact that a great deal of subsequent innovation takes place after such breakthrough innovations.

Uber serves as an illustrative case of breakthrough innovation. Uber has been one of the fastest-growing tech start-ups ever. Leveraging on GPS and smartphones, Uber made ride-hailing cashless, more reliable, and more convenient. Essentially, it engaged in innovation on several fronts – product innovation, disruptive innovation, business-model innovation, customer experience innovation, and cultural innovation. Subsequently, Uber encountered tough competition from similar companies, and it has struggled financially, in general and in certain regions, especially China. Nevertheless, the company has made a considerable effort to continue innovating in several respects, such as within the space of the self-driving car.

Uber's story began in Paris in 2008. Two friends, Travis Kalanick and Garrett Camp, were attending LeWeb, an annual tech conference that the *Economist* describes as "where revolutionaries gather to plot the future."[3] Rumor has it that the concept for Uber was born one winter night during the conference when the pair was unable to get a cab. Initially, the idea was for a timeshare limo service that could be ordered via an app. After the conference, the entrepreneurs went their separate ways, but when Camp returned to San Francisco, he remained fixated on the idea and bought the domain name UberCab.com.

An interesting simulation may be to consider how things might have developed had digital platforms on innovation been readily available when Uber was established in 2008 – today's advanced platforms for true data-driven innovation have only been available since 2018. In the early days, Valuer's platform could have given Uber start-up assistance ("1500 tools for start-ups" and access to the Academy). It could also have detected and followed the numerous attempts to plagiarize Uber's concept (the "Radar" service), which created many challenges for Uber, especially in Asia. Further-more, Uber used significant resources for developing autonomous driving solutions. The digital platforms could have given warnings or highlighted potential shortcuts. As Uber was a unicorn, CBInsights has followed its path. It could, for example, have helped optimize the timing of Uber's IPO.

If you were to create a similar breakthrough innovation, how would you utilize the digital platforms available today? Table 5.2 seeks to answer this question.

Use case D: "innovation accelerated by consolidation"

As another use case in which digital platforms on innovation may play a larger role in the future, we can look at innovation accelerated through consolidation. As we will see in Chapter 6, Tradeshift was initially very innovative within the space of e-invoicing and electronic payments. However, after it had reached beachhead status, it started

Table 5.2 Innovation-related services from digital platforms

Topical issue	First vendor response	Second vendor response
Finding an early stage start-up with a concept similar to Uber	Valuer's basic subscription	Preseries, Innovationscout, Findes, Innospot
Helping Uber as a start-up (in 2008)	What a Venture	Valuer's 1,500 tools and Academy
Assistance with patenting overviews, etc.	CBInsights	Specialized firms
Funding and matching, early stage	LetsVenture	Valuer
Funding, later stage	CBInsights	Funderbeam
Unicorn guidance, exposure, key metrics, etc.	CBInsights	Pitchbook
Radar service for other industries, e.g., automotive	Valuer.ai (at the micro level)	CBInsights (at the trend level)

actively acquiring companies, thereby growing its own industry-transformation power through consolidation.

Through the use of digital platforms, such strategies are likely to become easier to pursue. This is because digital platforms can help innovative start-ups identify attractive targets for upscaling.

On the future of digital platforms for data-driven innovation

Given the many vendors that have begun to target data-driven innovation in the last few years, we believe we are seeing a new industry emerge. This infant industry will develop according to certain business, including the following:

- Two-sided platforms will tend to grow faster than single-sided platforms because of double network effects. As we saw with the likes of Facebook, Amazon, Tencent, and Alibaba, network effects on both the supply and demand sides scale up very fast due to accelerated network effects. The same trend is likely to occur with the digital platforms, as more clients are attracted to platforms that offer more innovative content. At the same time, vendors and start-ups delivering innovative content to platforms are more likely to accelerate this trend as they see an increasing number of clients as the ultimate target.
- The sheer amount of data is key. For a serious digital platform, it is not enough to have just a few datapoints on a company in order to draw a conclusion or make a recommendation. In a world where value can be extracted from big data and in ways that you might not be able to envision, you need massive amounts of data on a large number of cases (start-ups) to generate the required insights. Therefore, getting to big data is important.
- Data quality is crucial. In order to be able to deliver quality output, it is not sufficient to have only thin data – you also need to include thick data to increase the quality of the output. This will often require the inclusion of qualitative data, which is a challenge because it is easier to collect and process quantitative data from, for example, existing databases. Qualitative data can be obtained through, for instance, qualitative interviews, the curation and cleaning of data, and trend analyses.
- Data handling is also key, especially when it comes to the analytical handling of data, as this activity can take different forms. It can, for example, be handled by advanced AI or by analysts. When advanced AI is used, the results will be free of subjective influence and based on, for example, skilled, deep learning. The use of analysts requires more resources, at least in the long run, and carries some degree of subjectivity. On the other hand, it may allow for approaches not foreseen in the specifications used in AI.
- The platform's geographical focus is important, as innovation is not confined to national or regional borders. A considerable number of platforms on start-ups are nation specific. This may make sense if the purpose is to track a geographically specific cohort of start-ups (e.g., for national grant purposes). However, the topic of this book is innovation, which is mostly global in nature. Therefore, we prioritize platforms that are global in nature.

Table 5.3 Key questions with regard to platform choice

Question	Platform characteristics
Is the platform designed and operated to deal with innovation as a one-sided or two-sided platform?	As a client, you are better positioned with a two-sided platform because your joining as a client (together with others) helps to strengthen the supply side.
What kind of quantitative data does the platform use?	There is a great deal of quantitative data in the era of big data. Look for platforms geared specifically toward data on innovation.
What kind of qualitative data does the platform use?	Qualitative data is resource demanding. Focus on the quality of the qualitative data.
How is data handled on the platform? Is there an active data-handling process?	Check whether data is handled by analysts or AI, or a combination of the two. With regard to AI, check whether it is advanced AI with deep learning or just light machine learning.
What is the geographical scope of the platform?	Platforms that have a truly global reach in their data on innovation have an edge.
Does one platform meet all of your requirements?	Client requirements may not be fulfilled by one platform at all times.

Few digital platforms can fulfil these criteria in the medium to long term. It will be interesting to see whether new platforms emerge or whether a wave of platform consolidation begins.

From the client's perspective, there is nothing awkward in subscribing to more than one platform. This is clear in some of the use cases discussed in this chapter. Based on the five business logics listed previously, Table 5.3 pairs client questions with the business logics behind the platforms.

Thus far, our focus has been on identifying and acquiring innovation, mostly from the perspective of corporate clients and mostly in relation to innovation in start-ups. The underlying rationale for an interest in innovation and for subscribing to a platform is often, if not always, to gain access to growth and, subsequently, see growth materialize.

However, do innovation and growth necessarily go hand in hand? If not, how can the interplay between innovation and growth be facilitated? These are the key questions addressed in Chapter 6. While we have been dealing with the close, causal relationship between start-ups and innovation, the next chapter takes innovation/start-ups as the starting point and progresses to exnovation/upscaling in the search for growth.

Key reflections

A   In Chapter 5, we addressed the important issue of how to use platforms that deliver data-driven innovation or innovation inputs. We also discussed some of the advantages of these platforms, such as their speed and relatively low costs, which improve efficiency, and the fact that they are better, which increases effectiveness. *Should data-driven innovation from the new platforms be viewed as a substi-*

tute for "old school innovation" or as a supplement? Alternatively, should data-driven innovation be viewed as just an instrumental means of measurement that should be part of "business as usual innovation"?

B   We presented a number of use cases ranging from "finding a needle in a haystack" innovation to "man on the moon" innovation. Given the increasing use of data-driven innovation, *is data-driven innovation relevant for all use cases or are there limits to how far one can reach with data-driven innovation acquired from the new platforms?*

C   Many of the new platforms deal with quantitative data. The justification for this approach seems obvious – it is cheaper and faster to establish a platform based on quantitative data, which may be gathered from existing databases. *What might you miss if you decide to deal only with data-driven innovation based on quantitative data? How can you assess the relative importance of thin data and thick data?*

Notes

1   CBInsights offers a broad range of services as do some of the other vendors. The characteristics given here aim to capture the main gist. For example, when the term "macro level" is used, it does not eo ipso disregard the possibility that the company may also deliver at the micro level from time to time.
2   Cf. PwC (2018). *SDG Reporting Challenge.* https://www.pwc.com/gx/en/services/sustainability/sustainable-development-goals/sdg-challenge-2019.html.
3   "The History of Uber: Uber Timeline." Accessed March 8, 2020. Cf. also: www.investopedia.com/articles/personal-finance/111015/story-uber.asp.

chapter 6

Crafting a growth strategy based on innovation

Access to attractive innovation does not necessarily mean that growth will automatically materialize. In general, a growth strategy encompasses the subset of both innovation and exnovation strategies. Exnovation is only vaguely addressed in the literature. For all practical purposes, it refers to scaling up an innovation by appropriating the value of that innovation in order to grow. In this context, execution skills become important key competencies as speed is often a critical factor in more than one sense. Exnovation can take a variety of forms. Where innovation deals with, for example, finding a needle in a haystack, exnovation involves scaling up or blitzscaling in an effective manner. In many instances, effectiveness is more important than efficiency. In other words, it is more important to see your growth strategy materialize than to ensure that all decisions are fully rational. When speed is of the essence, one cannot expect full rationality or the greatest possible efficiency, but effectiveness may be very high.

Access to attractive innovation does not necessarily mean that growth will automatically materialize. In general, a growth strategy encompasses the subset of both innovation and exnovation strategies as well as the interplay between them.

As innovation and exnovation are linguistically viewed as antonyms, exnovation may seem to be the opposite of innovation. However, a closer look at exnovation suggests that we are actually talking about complementary definitions. We have previously seen that innovation is closely tied to start-up entrepreneurship. Exnovation is similar to the upscaling of already achieved innovation.

With innovation, we are looking for something truly new. An innovation may be something we are unable to immediately comprehend, regardless of whether we are talking about "a needle in a haystack" innovation or a "man on the moon" innovation. With regard to innovation, subscribing to digital platforms is a new and effective tactic.

Exnovation is only vaguely addressed in the literature. For all practical purposes, it refers to scaling up an innovation by appropriating the value of the innovation in

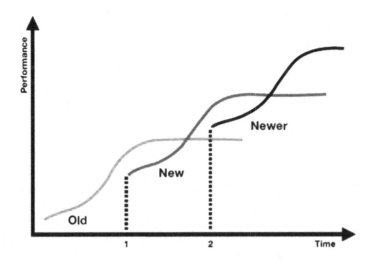

Figure 6.1 Optimizing growth through consecutive S-curves

order to grow. In this context, execution skills become important key competencies, as speed is often a critical factor in more than one sense. First, as part of a growth strategy, the steeper one can make the S-curve, the better. Second, speed is also important for ensuring that an S-curve reaching a point of diminishing growth can be taken over by the next S-curve, thereby ensuring corporate growth (see Figure 6.1). Third, speed is essential in order to achieve and retain market leadership. "Blitzscaling" is key in this regard.

As such, exnovation becomes a way to execute innovation in order to seek or rapidly accelerate growth. It occurs in addition to innovation or subsequent to innovation. In any case, the focus is on appropriating or extracting value from innovation in order to grow.

Exnovation can take a variety of forms. Where innovation deals with, for example, finding a needle in a haystack, exnovation involves scaling up or blitzscaling in an effective manner. In many instances, effectiveness is more important than efficiency. In other words, it is more important to see your growth strategy materialize than to ensure that all decisions are fully rational. When speed is of the essence, one cannot expect full rationality or the greatest possible efficiency, but effectiveness may be very high.

Instead of assuming that it is necessary to choose between innovation and exnovation, both should be in focus. In the long run, one is not sustainable without the other. The pursuit of growth strategies encompasses the dual subset of innovation and exnovation, and dealing with them ambidextrously gives additional benefits.

Table 6.1 summarizes some of the key components of a growth strategy that combines innovation and exnovation.

The failure of the strategic management literature to find the holy grail, the perfect recipe, or the evergreen formula for success has taught us to be cautious. Consequently, we provide some illustrative and widely different examples for inspirational purposes in the following subsections.

Crafting a growth strategy based on innovation 59

Table 6.1 Key components of a growth strategy

Subset of growth-strategy tactics	Innovation	Exnovation
Organizational fit	Start-ups	Scale-ups
Key competence	Entrepreneurship	Execution
Critical factor	Effectiveness	Speed
Main driver	Data driven, AI	Value driven, growth

Blitzscaling exnovation on top of innovation

Reid Hoffman and Chris Yeh co-authored the groundbreaking book *Blitzscaling*, which lists neither "innovation" nor "exnovation" in its index. Nevertheless, the book addresses exnovation from start to finish. According to the authors, it is key to (blitz-) scale the innovation or the start-up as quickly as possible and to reject a number of common governance principles in the process:[1]

> When should I start to blitzscale my company? . . . Doing so requires you to throw out many of the normal rules of business. It basically takes everything you thought you knew . . . and hurls it out of the window. Careful planning, cautious investment, courteous service, and a tightly controlled "burn rate" . . . may end up being tossed aside in favor of rapid guesstimates, ignoring angry customers, and inefficient capital expenditures. Why would you ever want to pursue such a risky and unintuitive course of action? In other words, speed.

One of the foremost examples given in the book is Airbnb. Airbnb received large amounts of funding, displayed exponential and speedy growth, and outcompeted others, taking many by surprise.

Aravind and McDonalds: frugal innovation combined with exnovation

Initially, Aravind's mission was to eliminate unnecessary blindness caused by cataracts in 15 million people in India. This was achieved by introducing innovations in relevant surgeries from start to finish. For example, Aurolab invented a US$8 lens (comparable to US$150 in the US and elsewhere). It also significantly reduced costs by bundling surgeries and by actively searching for multiple patients in order to achieve scale and efficiency. The orchestration of the surgeries was inspired, to some extent, by McDonald's. Widely viewed as the "inventor" of fast-food burgers, McDonald's industrialized mass-production and delivery systems.[2] Through a proprietary lab, alternative open hospital camps, and new surgery techniques (all involving innovation) as well as the utilization of the McDonald's format (equal to exnovation), Aravind established a hugely successful company. Similarly, Aurolab worked with an ambidextrous mindset to develop eye-related products that are now sold globally, thereby setting a new standard (i.e., exnovation).

Grundfos: discovering something new

Grundfos is a global leader in certain aspects of mechanical engineering in the clean-tech industry. It offers climate-friendly products, especially pumps, and services. Grundfos is often listed as one of the most innovative companies in the world, with one listing mentioning the following:[3]

> The idea of innovation at Grundfos is rooted and brought to life through different initiatives. Among these, the cooperation with start-ups, to welcome new disruptive ideas to Grundfos' corporate environment, and the opening, a few years ago, of an Innovation Center in the Silicon Valley to experiment and adopt new technologies.

Grundfos recognizes the value of data-driven innovation, and it buys access to innovations from start-ups found on a curated basis. Such systems allow it to keep track of industry-specific innovation being undertaken by start-ups, and to obtain access to formerly unknown ideas and methodologies normally placed under the heading of "innovation."

Through its search for innovation, Grundfos has also discovered new ways of producing products within the field of mechanical engineering. For instance, the company undertook a number of scouting activities focused on digital water and energy innovation. These activities identified an emerging cluster of blockchain applications related to its core business in water distribution, which triggered an additional focus within that space. More specifically, the blockchain search broadened the company's horizons with innovative solutions for decentralized utility management and distribution. It also identified solutions for internal processes such as an end-to-end smart supply chain platform. Therefore, what started as (acquired) innovation continued as exnovation, thereby supporting Grundfos's growth.

Another example of how innovation and exnovation may go hand in hand is Grundfos's search for innovation based on the Internet of Things (IoT). For instance, Grundfos's work with IoT in the context of smart cities calls for innovation and allows for exnovation, as the notion of smart cities allows Grundfos to find new ways to supply its climate solutions in larger quantities and in more efficient ways.

Last but not least, Grundfos is utilizing data-driven innovation to facilitate the achievement of the UN's SDG goals. In collaboration with NGOs, government agencies, and private companies, Grundfos runs the Lifelink project, which aims to improve access to water in the developing world. In addition to providing solar-powered pumps for the project, Grundfos developed the AQtap ATM, an automated water kiosk innovation that tracks water usage and allows locals to pay for water through water credits. The solution has already been installed in parts of Kenya, Uganda, Mozambique, and India as well as a number of other countries. By 2018, Grundfos and its project partners had provided more than 1.66 million people with access to clean water.

Nokia and Kodak: failing to work ambidextrously with growth

The Nokia case stands in direct contrast to the Grundfos case. As discussed previously, Grundfos established an innovation center in Silicon Valley. In the award-winning book, Ringtone, Doz and Wilson point out that if Nokia had developed a strong Silicon Valley presence, it would have been less likely to miss key innovations:

> A more determined and earlier move to follow the shift of the center of innovation to California – for argument's sake, an acquisition to bring in software architecture and project management skills, much as Technophone had earlier brought manufacturing skills – might have saved Nokia.[4]

In addition, Nokia's entire approach to innovation was tainted by the first wave of closed innovation, which focused more on speed than efficiency and effectiveness. Interestingly, Nokia's competencies, its ability to exnovate, and its production of standardized, high-quality handsets at a low cost are not discussed in the literature. As a market leader, Nokia did not seem sufficiently aware that its key competitive skill was not innovation but rather its world-class competencies in producing standardized handsets. However, what is the value of a world-class exnovation ability if a company is unable to access innovation and exnovate ambidextrously?

Kodak was once a leading exnovator with the analog camera. In contrast to Nokia, it was excellent at innovation. In fact, Kodak developed the digital camera. In this case, the company had the pieces needed to innovate (e.g., the digital camera) and it had adequate exnovative capabilities (from its experience with the analogue camera). However, it failed to ambidextrously deal with its growth strategies. In other words, Kodak failed to transfer its exnovation capabilities to the invention of the digital camera.

Siemens Gamesa: working ambidextrously with input from start-ups

Siemens Gamesa uses solutions built by start-ups to improve its own innovation in such areas as tracking and quality control within production. Furthermore, start-up scouting is used to identify solutions in data management, especially solutions able to systemize a wide array of unstructured data and sources.

This initiative is a side project aimed at potentially supporting or shortening the innovation cycle in the corporate technology department. Therefore, what started as innovation carried exnovation with it as a viable option for optimizing internal processes. The need to build entire new systems may therefore be eliminated if those systems are readily available from start-ups. This provides insights into how innovation and exnovation can share the same space and solve challenges within an organization if they are applied and carried out ambidextrously.

Tradeshift: blitzscaled innovation from a start-up

Tradeshift is a cloud-based business network and platform for supply chain payments, marketplaces, and applications. It achieved unicorn status in 2018 when a funding round, led by Goldman Sachs, raised US$250 million and the company was valued at US$1.1 billion.

Headquartered in San Francisco, Tradeshift serves more than 1.5 million companies and processes transactions measured in trillions of dollars, including transactions in the Chinese market. The company aims to connect all businesses on a global scale through e-transactions. Therefore, it has to simultaneously innovate and grow (exnovate).

Tradeshift seems to be a typical unicorn story. Its history demonstrates the presence of an ambidextrous mindset:

- In 2012, Tradeshift moved its headquarters from Copenhagen to San Francisco. In 2013, it opened an R&D center in Suzhou, China, and an office in London. Moreover, it added Latin American e-invoicing capabilities through a partnership with Invoiceware.
- In 2014, Tradeshift opened offices in Tokyo, Paris, and Munich. That same year, the EU Commission officially approved the Universal Business Language (UBL) data format – a format that Tradeshift supports – as eligible for inclusion in tenders from public administrations.
- In 2015, Tradeshift acquired the product information management company Merchantry, and launched e-procurement and supplier risk management solutions. In 2016, it acquired Hyper Travel and secured USD 75 million in Series C/D funding. In 2017, Tradeshift acquired IBX Business Network and launched Tradeshift Ada.
- In 2018, Tradeshift secured USD 250 million in Series E funding and launched Blockchain Payments as part of Tradeshift Pay. In December 2018, the company acquired Babelway, an online B2B integration platform. The acquisition added three new locations: Salt Lake City; Louvain-la-neuve, Belgium; and Cairo, Egypt.

This summary of Tradeshift's history demonstrates an ability to constantly innovate (i.e., continually develop the platform through proprietary and acquired innovation) and exnovate (as exemplified by the geographical expansion). Tradeshift has experienced extreme growth, including year-on-year revenue growth of 400%.

Broader perspectives

In recent years, big data and the development of AI have delivered several new ways of combining innovation and exnovation. Through the new data-driven innovation platforms, it is possible to answer such questions as:

a) *Can I find a needle in a haystack if I look for specific types of innovation or specialized start-ups on other continents?* (Yes)
b) *Is it possible to acquire information on whether certain start-ups are likely to survive?* (Yes)
c) *In addition to obtaining innovative input for my company, can I obtain guidance on how to standardize and upscale (i.e., exnovate)?* (Yes)
d) *Can I purchase an overview of likely game changers?* (Yes)

e) What about unicorns? Can I obtain information on upscaling companies that are likely to become unicorns as well as information on those companies after they become unicorns? (Yes)

f) From an investor's perspective, can I obtain specific information on start-ups with significant potential within, for example, the global cleantech industry as well as potential unicorns? (Yes)

The ability to answer such questions demonstrates that we are entering a new era characterized by greater transparency and the third wave of innovation/exnovation. Today, many decisions regarding innovation, exnovation, and related investments are based solely on selected knowledge. As such, they are often characterized by randomness and serendipity, and they are typically within the known universe. In contrast, real innovation with high exnovation potential can be found in the universe of the unknowns.

A bright idea from an R&D department or a perfect pitch from an entrepreneur may appeal not only to the rational part of a CEO's mindset but also to his or her emotions. In such cases and in the presence of analogue workflows, it is typically not standard operating procedure to use a data-driven approach to investigate whether a bright idea has already gained traction in other start-ups. Nor is it standard operating procedure to use AI to compare an entrepreneur's perfect pitch to the pitches of 100 or 1,000 similar start-ups. However, digital workflow efforts to do so lead to non-emotional responses to questions regarding the best ways to innovate and exnovate.

With the advent of data-driven, fast-track innovation platforms, making crucial decisions on how to innovate and exnovate has become quicker and easier. These platforms provide new ways of working ambidextrously with growth strategies that comprise both innovation and exnovation. In short, they allow for start-ups' activities to be rapidly upscaled.

Key reflections

A   In previous chapters, we deliberately focused on looking at the creation aspect of innovation or, in other words, how innovation is "generated." In Chapter 6, we take this one step further, as we look at growth as one way of appropriating value from innovation. *Is growth the only way in which you should try to appropriate value from innovation? Is growth the most important way to appropriate value from innovation?*

B   This chapter deals with the terminology and definitions behind innovation and exnovation (innovation scaled up in terms of market growth). It seems logical and intuitive to view this chronologically, such that innovation comes first, followed by exnovation. However, the chapter makes a plea for executing innovation and exnovation ambidextrously. *What are the advantages of dealing with exnovation as early as possible? When data-driven innovation is digital in nature, how much can you digitalize exnovation?*

C   Some examples of how (former) global blue-chip companies have fared with innovation and exnovation are provided. For example, Nokia was a world leader in

Crafting a growth strategy based on innovation

exnovation but disregarded disruptive innovation. Kodak was a global leader in innovation but failed to scale-up (exnovate) its valuable, proprietary innovation. *Are these extreme or typical cases? Which global blue-chip companies are good at executing innovation and exnovation ambidextrously over a longer span of time?*

Notes

1 Hoffman, R. and Yeh, C. (2019). *Blitzscaling: The Lightning-Fast Path to Building Massively Valuable Companies*. New York: Currency, p. 119f.
2 Andersen, M. M. and Poulfelt, F. (2014). *Beyond Strategy: The Impact of Next Generation Companies*. New York: Routledge, pp. 88–92.
3 A blog post on innovation is available on the Valuar.ai website. This quote is from "100 most innovative large companies." Notably, Grundfos currently has no innovation center in Silicon Valley.
4 Doz, Y. L. and Wilson, K. (2018). *Ringtone: Exploring the Rise and Fall of Nokia*. Oxford: Oxford University Press, p. 156.

chapter 7

Heading toward a digital innovation economy

The world has a hunger for growth and one of the most important vehicles for growth is innovation. With the new digital platforms from which data-driven innovation can be extracted, major parts of analogue workflows will be replaced by digital workflows. Subsequently, the field of innovation will eventually reap the benefits of digitalization. In a world where start-up creativity and activity have never been higher, the hope that the innovation behind a start-up will become something big remains as strong as ever. Entrepreneurs still dream of becoming rich. Likewise, highly skilled engineers who work in R&D departments hope to "invent" the new innovation. The new digital platforms allow for many entrepreneurs and R&D departments to collaborate and, thereby, reach a higher level faster. With the advent of the new digital platforms, (successful) innovation becomes more predictable, purpose-driven, profitable, and probable.

Connectivity and the digital innovation economy

We are living in a digital world. In fact, the digitization of Western societies started more than 50 years ago. Notably, the telecommunications networks in some countries were already fully digitized in the 1970s. Broadband is now widely available in most countries, often at relatively low prices, which incentivizes customers to substitute physically demanding work patterns with virtual work processes and remote/external access to knowledge. In this context, the important thing is not to focus on the technology as such but on whether adequate applications are in place to deliver attractive products and services.

Interestingly, we can distinguish between an old or existing innovation economy and the new digital innovation economy. In the existing economy, we often measure the degree of innovation in a given country in terms of R&D's percentage of GDP. We assume that more innovation takes place if this ratio is high. Moreover, we assume that if the ratio is high, then growth will also be higher as the level of innovation is higher. These assumptions may not be correct. They were more likely to be correct in the "old" economy where innovation was tied to traditional R&D activities that heavily relied on analogue workflows. Somewhat surprisingly, the field of innovation has been characterized by analogue workflows and, at least at the micro level, by a failure to utilize data-driven innovation and AI.

However, several preconditions for these causalities are beginning to change and create a new picture. One salient discovery is the fact that innovation is not just created in large, incumbent R&D organizations. It may be even stronger in start-up communities. This means that assessing, for example, a country's level of innovation will become far more difficult, as it is no longer just a matter of measuring R&D's percentage of a national, regional, or global GDP.

In the new innovation economy, it is more a matter of quickly finding pieces of innovation that are already developed. Therefore, we need a more advanced means of measurement in order to understand these new processes. One such measure may be the degree of connectivity between start-ups and corporations, and connectivity within these sub-segments. At the micro level, we return to the importance of data-driven innovation from digital platforms as their footprints seem to grow quickly and in accordance with the shift from primarily analogue to primarily digital innovation workflows.

The world has a hunger for growth, and one of the most important vehicles for growth is innovation. With the new digital platforms from which data-driven innovation can be extracted, major parts of analogue workflows will be substituted by digital workflows. Subsequently, the field of innovation will eventually reap the benefits of digitalization.

The combination of the SaaS business model with cloud-based software

One of the underlying global trends that attracts a great deal of attention during times of crisis is the SaaS model. Salesforce probably deserves credit in this contest – not for inventing the model but for scaling it up. Given its excellence in exnovation, Salesforce paved the way for an attractive business model with a combination of cloud-based software and the SaaS business model. It is simply a subscription payment model with recurrent revenue streams.

The attractiveness of this model in the context of data-driven innovation is based on the following key characteristics:

- When a platform is cloud-based, the software and the intelligence attached to that platform means that customers can share solutions and reap the benefits of previous learning. This is similar to a translation algorithm becoming smarter each time it is used. In this case, a digital platform becomes better at predicting matches for innovation partnerships based on previous results.
- There are no physical installations at the customers' premises. Instead, virtual access to the common platform is provided. This not only lowers the relative costs but also reduces delivery times. This is key to understanding why data-driven innovation from digital platforms based on the SaaS model helps increase both the efficiency and effectiveness of innovation processes.

- In contrast to the traditional software-licensing model, customers avoid huge upfront license fees and the costs associated with installation and integration. This is attractive from a liquidity perspective. The vendor is also satisfied because a long(er) recurrent revenue stream is secured. The capital markets reward the SaaS model with far higher multiples than those applicable to conventional business models. As a considerable number of digital platforms are based on SaaS, many of them have a strong staying power. Conversely, digital platforms based on consulting are subjected to higher volatility when it comes to revenue streams.
- The use of the cloud for platforms facilitates the proliferation of network effects on both the demand side (i.e., customers, notably large corporations) and the supply side (i.e., start-ups). When a certain critical mass has been reached, the future will be very bright for a number of these digital platforms.

Many of the platforms displayed in Chapter 4 rely on a cloud-based SaaS business model. As such, they can deliver data-driven innovation in electronic, virtual environments at comparatively low costs. This modality for innovation is interesting in general and, in particular, in the COVID-19 scenario. In the aftermath of the COVID-19 crisis, several additional advantages are evident. In particular, the restrictions on travel and physical cooperation resulting from COVID-19 have served as a catalyst for digital workflows in general and have created a spillover effect on existing innovation processes that were previously closely tied to analogue processes.

The challenges faced by analogue innovation workflows

In a normal innovation scenario in a conventional mode, the key characteristics typically include the following:

- Innovation takes place through a physical presence in the company's R&D department.
- Meetings are almost always held in person (including external meetings).
- Some prestige is attached to conducting meetings on other continents, meaning that overseas travel is "necessary" in order to attend professional conferences and/or reach out to the best experts.
- The identification of bits and pieces of innovation from the start-up ecosystem is random and almost completely based on subjectivity.

Yet again, we see a serious dilemma with some of the existing modes of innovation, as we briefly discussed in Chapter 3. If the chosen modality is closed, then effectiveness falls to an unacceptably low level. If an open modality is chosen, then the efficiency monitor often displays long incubation periods as well as high costs. In this context, conventional companies would struggle to strike a balance and find a path to a decent compromise between efficiency and effectiveness.

The new heroes reject such engagement with suboptimal solutions involving compromises. Instead, they focus on simultaneously achieving the greatest effectiveness and the greatest efficiency. This brings us to data-driven innovation and the transformation to digital workflows, which may emerge as a cornerstone of the best modality for innovation in the future.

Data-driven bailout: the four Ps

As we previously eluded to, the utilization of data-driven innovation carries the advantages of the four Ps. More precisely, data-driven innovation makes both innovation and growth more predictable, purpose driven, profitable, and probable.

As we have seen, conventional innovation in closed environments does not suffice. Open innovation is associated with serendipity. This relates to, for example, the randomness in finding other organizations with which to collaborate. A company seeking external partners for innovation development purposes will often look to a nearby university or research institution for external inputs. In the best cases, companies attempt to domicile themselves in relevant clusters. For instance, many companies in the pharmaceutical industry locate themselves in the greater areas of Boston or Copenhagen. However, data-driven innovation through platforms makes finding the right pieces of innovation much more **predictable** because the customers of these platforms are able to quickly search a great deal of relevant data on innovation.

For example, searches for companies facing challenges based on COVID-19 could generate a wide array of results:

- Industries relying on unskilled human workforce could look into automation (to keep factories/product running).
- A surge in delivery services and solutions is one result of the crisis. This has led to higher profitability as well as customer interest in previously unseen formats for subscriptions of physical goods and deliveries of essential goods.
- Waste and re-use has also experienced an increased focus, especially when it comes to food waste and crops that are not being used in the hospitality business.
- Spaces and resources that are not being used during the lockdown can be utilized in other ways or refurbished.
- There is a focus on remote work and collaboration (e.g., task management).

Customers can also quickly ensure that these data are curated and, thereby, ready to use. As such, they can learn about innovations of which they were previously unaware.

Imagine that a pharmaceutical company would like to develop a COVID-19 vaccine internally. Access to the new platforms would be key, as the development would be more **purpose driven**. There are many ways in which data-driven innovation can help companies and societies under the difficult circumstances created by COVID-19. Table 7.1 illustrates just a few. It also demonstrates that even the platforms themselves are driven toward the higher purpose of helping out in crisis situations.

Successful innovation is not only about creating something new in a closed environment or about launching data-mining exercises on new platforms. The path to **profitability** requires a combination of innovation and exnovation, as discussed in Chapter 6. Long before vaccines for COVID-19 become available, effective medications

Table 7.1 Example of purpose-driven innovation from the platforms (May 2020)

CBInsights, examples	Valuer.ai, examples
Picking the most valuable and value-increasing pharma and health tech companies	Identifying start-ups working with vaccines in general and for COVID-19 in particular
The impact of COVID-19 on venture investments in start-up companies	Ranking start-ups in the affected ecosystems on how well they handle the challenges associated with COVID-19
How AI is predicting drug efficacy, discovering new antibiotics, and accelerating the diagnosis and treatment of COVID-19	Start-ups working with relevant COVID-19 subjects, especially speed and agility in moving from testing to consumer phases; new technologies and solutions that are being applied to change the ways that vaccines, treatment, and social distancing are managed

or other treatments may be developed. As we have seen time and again, innovation is almost useless without exnovation. Successful innovation is not just about innovators innovating. It is also about customers adopting the innovation and about product-market fit. In general, millions of research hours can be saved by substituting proprietary innovation with data-driven innovation from the new platforms, thereby moving toward higher profitability.

In a world where start-up creativity and activity have never been higher, the American dream that the innovation behind a start-up will become something big remains as strong as ever. Entrepreneurs still dream of becoming rich. Likewise, highly skilled engineers who have worked for years in the R&D department of, for example, pharmaceutical companies still hope to "invent" the new blockbuster drug that will give them recognition and huge bonuses. The new digital platforms allow for many entrepreneurs and R&D departments to collaborate and thus reach a higher level faster. This makes successful innovation more **probable**.

With the advent of the new digital platforms, (successful) innovation becomes more predictable, purpose-driven, profitable, and probable. Dealing with innovation in a data-driven way may give society at least two advantages during times of crisis. First, data-driven innovation may be one important path to addressing crises. In relation to COVID-19, the best way might be to find the right vaccines. Second, data-driven innovation combined with exnovation can accelerate growth, which is often needed in the aftermath of not only financial crises (e.g., 2008–2009) but other crises (e.g., COVID-19), which create challenges that last for years.

The 2008 financial crisis led to a focus on profitability at the expense of growth, such that levels of innovation shrank. The COVID-19 crisis and its aftermath will most likely lead to a stronger focus on growth, perhaps at the expense of almost everything else, including profitability. Data-driven innovation is one of the most important keys of returning growth to historic levels without sacrificing profitability. In this context, digital platforms offer possibilities to combine such strongholds with imponderables like the UN's SDG goals.

A client of a data-driven platform can adhere fully or partially to an SDG as a starting point. They may look for start-ups that can help bridge the gaps among their organization's current business model, competitive advantages, and the need to address strategically chosen SDGs.

Not only is utilizing a data-driven approach to align SDGs with the core business a more efficient method of innovation but it may also be the only way to identify previously unknown similarities or to match business models in order to resolve problems. Many organizations have a traditional focus of improving their current product portfolio to improve profitability without investigating opportunities outside their current market boundaries. By identifying SDGs that are strategically important or relevant for an organization without investigating opportunities for profitability, an organization can open up for an explorative search within previously unaddressed markets.

An organization can utilize data-driven innovation from a digital platform to combine its current capabilities and organizational strengths with a chosen SDG and then apply a digital platform. At that point, the matching algorithms can locate start-ups that are both relevant for the organization's capabilities and applicable to the chosen SDG. This creates new opportunities in which organizations can expand into new markets based on sustainability requirements, such as the SDGs, thereby, enable sustainable business models, and actively solve the challenges that the SDGs represent.

An AI platform is useful and sometimes necessary when matching an organization with SDG challenges requires the representation of several goals in the same results. When combining a corporation's profile with potential market opportunities and one or more SDGs, considering the most optimal parameters requires a process that is able to handle multiple inputs and variables. Thus, a corporation trying to match both profits and SDG challenges must use a technique that can process a variety of parameters. An AI-powered digital platform is necessary to efficiently utilize data-driven innovation and predict the most optimal outcome/match.

An emerging mode of innovation and an emerging innovation economy

A number of digital platforms have emerged in recent years from which data-driven innovation can be created and value can be appropriated. The present situation as well as the many examples we have discussed illustrate that a new and much stronger approach to innovation is gradually replacing conventional modes of innovation. As a result, many analogue workflows will be replaced by digital workflows in the near future.

If this development gathers speed, the network effects will culminate in a new digital innovation economy. In this economy, the optimization of the connectivity within start-ups and corporations and between start-ups and corporations will continue to

Table 7.2 Three modalities for the development of innovation

Closed innovation	Open innovation	Data-driven innovation
Centralized, inward innovation	Partially external, collaborative innovation	Innovation from platforms based on network effects
Traditional research culture	"Oxford meeting" culture	Growth/exnovation culture
Low effectiveness	Low efficiency	Higher effectiveness and higher efficiency
Serendipity-driven usage of start-ups	Use of start-ups (e.g., through accelerator programs)	Strongest possible scouting, curating, and usage of start-ups
Analogue workflow	Mainly analogue workflow	Mainly digital workflow

accelerate. This emerging innovation economy originates from the trend depicted across the three innovation modalities illustrated in Table 7.2.

The new trend of data-driven innovation from digital platforms will be important for future success on the micro level, as data-driven innovation goes hand in hand with both efficiency and effectiveness. This fosters higher growth and, therefore, an attractive framework for business-model success.

Similarly, data-driven innovation will be key for future success at the macro level. Data-driven innovation and the widespread application of digital and AI-based workflows may kickstart an innovation economy at the aggregate level. This innovation economy is likely to be characterized by successful growth, higher GDP numbers, and lower R&D budgets. It may also support higher purposes, such as the UN's SDG goals, and better handling of certain crises.

Key reflections

A   Chapter 7 deals with the advantages of the SaaS model relating to the new digital platforms, most of which are based on this business model. *Thinking of the future, what is the likely business penetration of these platforms when data-driven innovation increases efficiency and effectiveness, and the underlying platforms are based on a business model with strong and increasing recurrent revenue streams?*

B   The chapter deals with the four Ps that characterize data-driven innovation from the new platforms: **p**redictable, **p**urpose-driven, **p**rofitable, and **p**robable. *Can weaknesses be identified that will lower the penetration of these new platforms (e.g., the culture and conservatism of the R&D departments of large corporations or NIH syndrome)?*

C   With regard to the COVID-19 crisis and other crises, data-driven innovation from cloud-based platforms based on the SaaS business model seems to be on the winning team. In 2020, companies that are digital, cloud-based, and able to replace physical workflows with digital, virtual solutions are growing quickly. *Can data-driven innovation from these new platforms help reboot growth? Can data-driven innovation help mitigate the effects of COVID-19?*

Heading toward a digital innovation economy

Appendix A: Two tools for dealing with innovation and growth

1 Tools related to the invincible company

The first type of tool was developed by Alexander Osterwalder and his team and is best known under the heading "Business Model Canvas." Furthermore, the notion of "The Business Model Portfolio" was recently introduced in the book *The Invincible Company*.[1] The strength of these highly successful tools results from rebuilding the corporate company using existing and readily available resources, thereby achieving business-model innovation. However, in Osterwalder's new book (*The Invincible Company*), a distinction is made between "explore" and "exploit" (see Table A.1).

"Explore" is focused on everything we normally relate to innovation, while "exploit" is focused on execution. With this toolbox, the idea under "explore" is to launch and experiment with many projects, some of which will partly succeed. Subsequently, "exploit" takes over and executes successful ideas and ensures the successful scaling

Table A.1 Key differentials between "explore" and "exploit"[1]

	Explore	*Exploit*
Focus	Growth	Cost-cutting/efficiency
Investment philosophy	Venture-like	Stock-market-like
Culture	Experimentation	Linear execution/"failure not an option"
People/skills	Exploration	Detail/rigor

1 This distinction is widely known from the literature, e.g., March, 1991.

in order to uncover a successful growth engine. As such, there are some general similarities between how Osterwalder uses "explore" and "exploit," and how we used the terms "innovation" and "exnovation" in Chapter 6. However, please note that Osterwalder's use of "explore" and "exploit" differs from March's original definitions.

Some of the mantras referred to are:

> It takes 250 projects to produce 1 Nespresso growth engine.
>
> —Osterwalder

> You can't pick the winner without investing in the losers.
>
> —Osterwalder

> Failure and inventions are inseparable twins.
>
> —Jeffrey Bezos, Amazon

This is somewhat similar to experiences in the venture capital business regarding early stage investments. This philosophy is represented in the "Funnel" tool, displayed in Figure A.1.[2]

We start with innovative projects, ideas, and start-ups in the lower left-hand corner. These are high-risk projects with low expectations for financial returns. Throughout the explorative phase, the number of projects are reduced. Those that remain are moved on to the "exploit" phase for execution with the ultimate aim of having one new growth engine.

The invincible company masters dealing with innovation ("explore") and exnovation ("exploit") ambidextrously.[3] However, as a prerequisite, the invincible company must always have an attractive innovation pipeline.

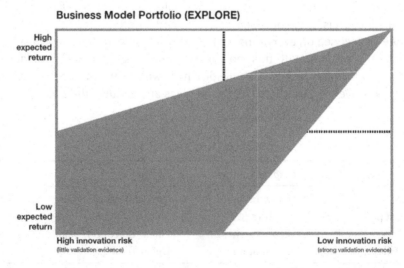

Business Model Portfolio (EXPLORE)

High expected return

Low expected return

High innovation risk
(little validation evidence)

Low innovation risk
(strong validation evidence)

Figure A.1 The funnel tool

Tools for dealing with innovation and growth

Alexander Osterwalder and his team, notably Michael Wilkens, already provide innovation services based on a coherent toolbox for business model innovation. Given the toolbox for the invincible company, it seems like more attention will now be paid to innovation not only at the generic level of business models but also at the micro level with the innovation funnel tool.

2 Tools related to blitzscaling

The second type of tool is delivered by Reid Hoffman and Chris Yeh and include a telling subtitle for their master book, *Blitzscaling*[4] – "The lightning-fast path to building massively valuable companies." Massively valuable companies are precisely what Hoffman has managed to help create in numerous instances through co-founderships or investments (e.g., PayPal, LinkedIn, Airbnb, Dropbox, Facebook, Instagram, Red Hat, and Tumblr).

The primary focus of the blitzscaling toolbox is not corporate incumbents but rather the subsequent phases of potent start-ups. Where existing corporations prioritize efficiency, blitzscalers adhere to a recipe in which speed is the most important ingredient.

Established corporations have only a few potential advantages when it comes to blitzscaling:

- Large corporations already have some scale and can therefore rely on their existing economies of scale (e.g., Amazon).
- Existing corporations are able to make multiple, iterative attempts to blitzscale. A prime example was seen in the first and second attempt of Microsoft Windows to copy Apple's Macintosh. The third attempt (i.e., Windows 95) was successful.
- Existing corporations can be patient. Google and Facebook are prime examples in this regard.
- Large corporations have the resources and the power needed to pursue M&A deals (e.g., Google's acquisition of Android and Facebook's acquisition of WhatsApp).

Large corporations also have disadvantages when it comes to blitzscaling. Among these are incentives to favour cautious expansion rather than aggressive blitzscaling, a general unwillingness to stage investments, and frequent pressures to deliver short-term results.

Start-ups often have some advantages relative to large corporations. For instance, some of them are able to attract much more capital than corporations, at least when it comes to serial rounds of capital injections. While there are many start-up hubs globally, there are only a few scale-up hubs.

What do start-ups need to attract capital? Obviously, talent needs to be in place. Additional criteria are listed in Table A.2.

Just like Osterwalder and his team created Strategyzer out of the Business Model Canvas and The Invincible Company, Reid Hofmann's co-author, Chris Yeh established the Global Scaling Academy together with Jeff Abbott to offer consulting on blitzscaling. One part of their general toolbox revolves around the requirements for blitzscaling

Table A.2 Requirements for blitzscaling[1]

Key requirements for blitzscaling
Big market
Massive distribution
High gross margins
Network effects
Product-market fit
Operational scalability

1 From a presentation given by Chris Yeh to start-ups and corporations in Copenhagen January 14, 2019.

Table A.3 A comparison of the invincibility and blitzscaling approaches

	The invincible company	Blitzscaling	Comments
Focus	An (incumbent) corporation	A potent start-up or SME	From big to small, from small to big
Many/few	Many projects, few successes	Few/one, spreading to others	Many versus few
Methodology	Innovation funnel	Blitzscaling talent	Funnel versus talent
Implied techniques	Efficiency Systematic and predictable	Speed	Both in the face of uncertainty
Use of platforms	Discovery and curation	Discovery and curation	Equal types of input needed
Ultimate goal	The invincible company	The massively valuable company	Almost equal, i.e., invincible versus valuable

shown in Table A.2. Other parts of the toolbox relate to other parts of blitzscaling that they are licensed to use. When it comes to data-driven innovation, the Global Scaling Academy has developed a suite of digital advisory services together with Valuer.ai in order to support digital workflows in innovation processes.

A comparison of real innovation in the approach to the invincible company and in a blitzscaling scenario is provided in Table A.3.

Both approaches have a similar core goal: to create an invincible or massively valuable corporation. They are almost identical with regard to the possibility of utilizing the platforms for data-driven innovation. However, their methodologies, strategies, and tactics are notably different. One approach deals with a funnel of innovation based on many start-up-like activities and an analytical approach based on efficiency, while the other is occupied with scaling up speedily, often at the expense of efficiency.

Tools for dealing with innovation and growth

Notes

1 This is, at the time of writing this book, the title of a forthcoming book by Alexander Osterwalder and his team, which is scheduled to be published by Wiley in 2020.
2 Ibid.
3 This is from a blog conversation between Osterwalder and Steve Blank, February 14, 2019, cf. "StratChat with Steve Blank: Startups vs Big Corporations – Who Will Win the Disruption War?", *Strategyzer*, https://www.strategyzer.com/blog/posts/2019/2/14/stratchat-with-steve-blank-startups-vs-big-corporations-who-will-win-the-disruption-war.
4 Hoffman, R. and Yeh, C. (2018). *Blitzscaling: The Lightning-Fast Path to Building Massively Valuable Companies*. New York: Currency.

Appendix B: The inner workings of a digital platform for data-driven innovation

Platforms for data-driven innovation are built in various ways. In this appendix, we provide an illustrative example. The descriptions do not purport to illustrate what happens on the supply and demand sides. Instead, the focus is on the inner workings of data with the platform and on the client's perspective.

The starting point occurs in two places. From the client's perspective, the mission is set up by feeding criteria into the platform using a selection of parameters purposefully designed to cover input required for the selection and matching process. For the start-up, the starting point occurs when it enters the system through one of the sourcing mechanisms applied by the platform.

There is a logical chronology:

- Data sourcing

  - The platform is fed data from a wide array of sources ranging from API connections to crowdsourcing networks. These sources are all fed into the temporary database, which holds new entries until they have been validated, processed, and (potentially) mapped to previous entries.

- Data enrichment

  - The enrichment process feeds the data through several mechanisms designed to unify and validate it, and to prepare it for matching with clients. The data, which contains a significant amount of text, is translated into English for NLP processes. It is enriched by adding and combining data to previous entries, which are continuously updated (a ranking system prioritizes entries based on their sources, and data overwriting only occurs when there is a high degree of trust).

- Data processing

  - During the data processing, the data is systemized and placed in relation to the larger dataset. Data entries are clustered based on unsupervised learning, and client criteria are processed and vectorized to prepare them for positioning in the database for the matching process.

- Applying the data

  - The vectorized search definition is projected (placed) into the clusters generated from the start-up data, which identifies a point in close vicinity to the relevant start-up clusters. This process highlights start-ups that may be of value for client evaluations and, later, selection. After client feedback is received, the vector is adjusted and projected into clusters with even greater relevance. This process is repeated until a satisfactory fit is achieved.

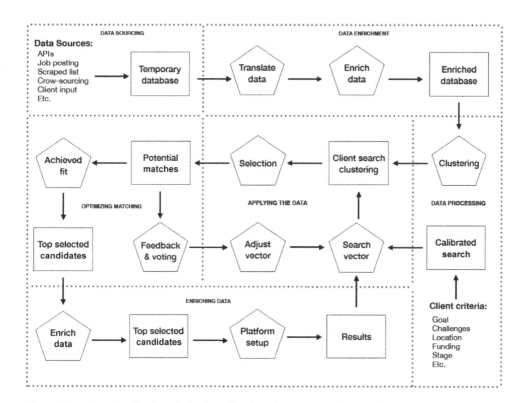

Figure B.1 Example of a digital platform for data-driven innovation

The inner workings of a digital platform

- Optimized matching

  - The clusters identified as potentially relevant for the calibrated search are chosen for client evaluation, which results in the selection of key examples of each cluster for presentation and feedback through the platform. The feedback on the clusters acts as reinforcement learning for the vectorizing, which feeds into the adjustment of the projection. This process is repeated until the clusters reach a satisfactory level of fit. The clusters are then processed to develop a listing of the highest-ranking candidates for later enrichment.

- Data enrichment

  - The enrichment process fills out and uniforms qualified data for the highest-ranking candidates in order to enable final selection for presentation. The platform then summarizes and presents the results to the clients and feeds the selection results back into the training algorithm for use in future matching processes.

The inner workings of a digital platform

Appendix C: Platform price comparison

This appendix purports to replicate the prices for two vendors. The prices come in a reduced version and are only for the two candidates in the paired comparison.[1]

Table C.1 Pricelist from CBInsights

Enterprise $2M+	Executive	Insider	Advisor	Prime	Answer	Analytics
Annual fee (USD)	$2,65,000	$1,99,000	$1,28,640	$99,700	$79,900	$59,690
plus one time 30-day onboarding fee*						
Platform and intelligence						
Customer success manager	✓	✓	✓	✓	✓	✓
Private and public company and market data	✓	✓	✓	✓	✓	✓
Financing, MSA, IPOs, performance metrics						
Mobile app	✓	✓	✓	✓	✓	✓
Client-only newsletter	✓	✓	✓	✓	✓	✓
Collaboration and workflow	✓	✓	✓	✓	✓	✓
Collections notes and stories						
Mosaic scores	✓	✓	✓	✓	✓	
Expert collections	✓	✓	✓	✓	✓	
Market map maker	✓	✓	✓	✓		

(Continued)

Table C.1 (Continued)

Enterprise $2M+	Executive	Insider	Advisor	Prime	Answer	Analytics
Enhanced data						
Patent analytics	✓	✓	✓	✓		
Earnings transcript analytics	✓	✓	✓	✓		
Business relationships	✓	✓	✓	✓		
Market sizings	✓	✓	✓			
Unlimited research access						
Expert intelligence	✓	✓	✓			
Scout**						
Scout missions		unlimited				
Smut select minions	unlimited					
Community						
Conferences and events	discount	discount	discount	discount	discount	discount

	STANDARD $12,000 PER MONTH *unlimited users*	PREMIUM $24,000 PER MONTH *unlimited users*
PLATFORM	• Define one mission every 90d • AI + agent-based initial search • Curation of top 25 matches by our analysts and expert network • Complete online report for top 25 matches	**All features of the standard plan PLUS:** • Quarterly customer success mission review

RESEARCH	• Printed research report for top ten matches
	• Relevance-based matching only
	• Blitzscalability assessment for ten matches
		All the features of the standard plan, PLUS:
		• Second level corporate needs assessment interviews
		• Second level startup assessments for advanced matching by:
		  • Problem / Solution fit
		  • Team / Operational scalability
		  • Cultural fit
		  • Technology robustness
		  • Functional area scalability challenges
		  • Blitzscalability
		  • VC Investment review: how VCs would look at the startup

ACADEMY	• Step-by-step video tutorials to getting the most out of Valuer.ai platform • Self-guided online content on: • Working effectively with startups • Corporate Innovation best practices	**All the features of the standard plan, PLUS:** • Expanded content library featuring top subject matter experts and thought leaders. • Advanced scaling and innovation-related educational content for corporate leaders • Specialized content for start-ups you invite • Personalized explanations of curated research

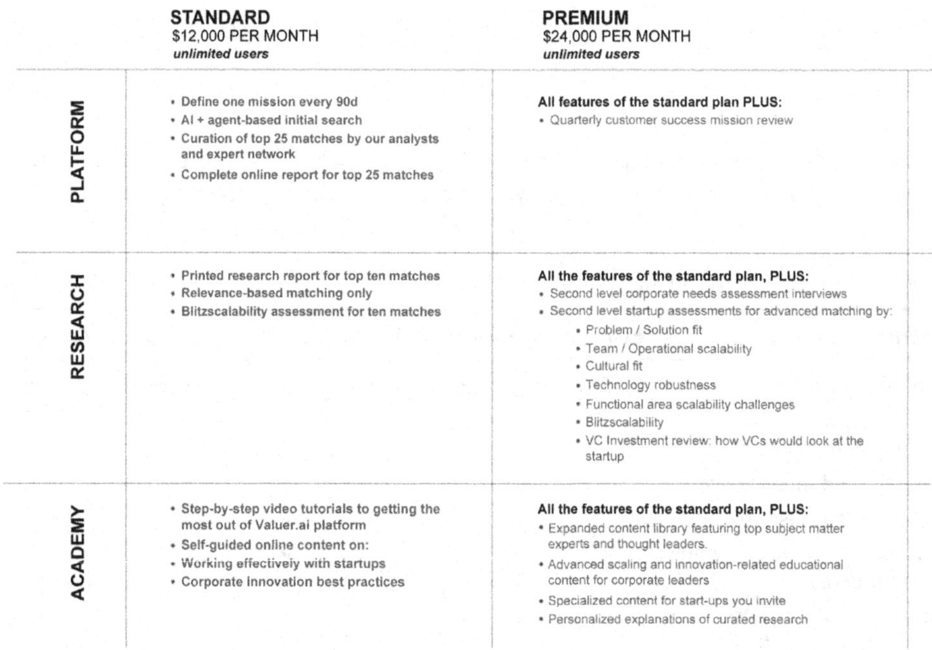

Figure C.1 Pricelist from Valuer.ai

Note

1 From their websites and additional interviews.

Appendix D: Semantics of "innovation," "data-driven," "AI," and "growth"

Innovation

"Innovation" comes from the Latin *innovationem*, a noun of action that comes from *innovare*. *The Etymology Dictionary* states that *innovare* dates back to at least around year 1500 and stems from the Latin *innovatus*, a past participle of *innovare* – to renew or change" – from *in-* "into" and *novus* "new." The central meaning of innovation relates to renewal. Despite the etymological and intuitive meaning, we find more than 40 definitions in the literature.

Innovation is the opposite of exnovation, which implies a stoppage of renewal. This makes sense if there is the desire to standardize things (e.g., for scaling purposes).

There are a number of different types of innovation.

In Geoffrey A. Moore's book *Dealing with Darwin: How Great Companies Innovate at Every Phase of Their Evolution*,[1] innovation is considered in the context of the category life cycle, with category being the product or service term used by customers that distinguishes what they are buying. In this context, Moore defines the following innovation types:

- Disruptive;
- Application;
- Product;
- Platform;
- Line extension;
- Enhancement;

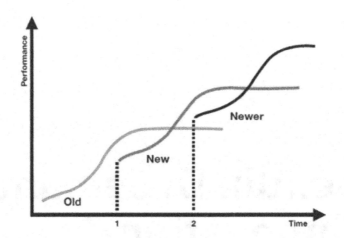

Figure D.1 Types of innovation

- Marketing;
- Experiential;
- Value engineering;
- Integration;
- Process;
- Value migration;
- Organic; and
- Acquisition.

Doblin[2] suggests 10 types of innovation from industry patterns. These include innovation in:

- Business models;
- Networking;
- Enabling processes;
- Core processes;
- Product performance;
- Product systems;
- Services;
- Channels;
- Brands; and
- Customer experiences.

Moreover, distinctions are made in the literature with regard to the change impact or scope of innovation, leading to the following common types of innovation:

- Incremental;
- Radical (or breakthrough); and
- Transformational (or disruptive).

"innovation," "data-driven," "AI," "growth"

In an interesting twist, the word "innovation" also seems to have transformed into shorthand for "anything new" and/or "anything good." Google's Ngram database of word use finds that not only has "innovation" become a bigger deal than "invention" but also that total mentions have reached an all-time high.

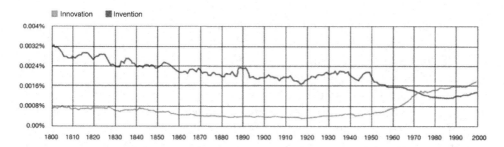

Figure D.2 Innovation versus invention in %

This development has continued in recent years as "innovation" continues to outnumber "invention." Moreover, we see some interesting spikes during the initial phase of the last financial crisis.

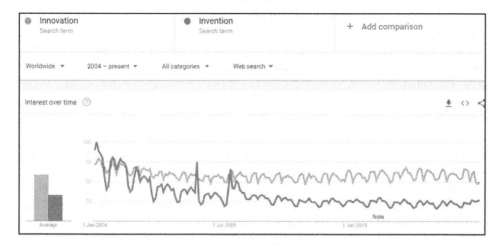

Figure D.3 Innovation versus invention
Source: Google Trends (May 19, 2020).

Moreover, a "lack of innovation" has become the easiest way to explain everything from slow job growth to Nokia's derouting. This is not a new phenomenon. Even Machiavelli's *The Prince*, which experts suggest was written in 1513, deals with these two sides of the coin (i.e., good and bad). Consider the following requote of the passage we quoted in the foreword:[3]

It ought to be remembered that there is nothing more difficult to take in hand, more perilous to conduct, or more uncertain in its success, than to take the lead in

"innovation," "data-driven," "AI," "growth" 85

the introduction of a new order of things. Because the innovator has for enemies all those who have done well under the old conditions and lukewarm defenders in those who may do well under the new. This coolness arises partly from fear of the opponents, who have the laws on their side, and partly from the incredulity of men.

Military people have interpreted this as dealing with different kinds of military motivations – to win or lose, individual motivations, status, prestige, and dominance. In the US, the military has continually discussed whether Machiavelli was right or wrong (e.g., Captain Brad de Wee, January 18, 2016, "Was Machiavelli Right about Innovation?").[4]

Despite this discussion, in this book we ignore the power struggle that occurs on the cultural side when innovators encounter resistance "in the introduction of a new order of things." Five hundred years after Machiavelli's work this is still relevant. First, innovation often meets with resistance, at least in large, conventional organizations. Second, and perhaps more importantly given the focus of this book, data-driven innovation may well represent "a new order of things." Therefore, according to Machiavelli, a breakthrough is not necessarily automatic.

Innovation was not always a positive term. In the seventeenth century, "innovators" did not get accolades. Instead, their ears were cut off. (The following is based on the work of Emma Green, a staff writer at *The Atlantic*.)[5] The irony behind the fact that innovation is now a buzzword is that, originally, "innovation" was not a compliment – it was an accusation.

In fact, shouts of "Innovator!" used to be akin to charges of heresy. As with any question of intellectual history, the path of innovation over the centuries is complicated. Canadian historian Benoît Godin has done extensive research on the topic. Although we greatly oversimplify his work here, a few moments in the strange history of the term "innovation" seem particularly striking.

According to Godin, innovation is the most recent incarnation of previously used terms like *imitation* and *invention*. When "novation" first appeared in thirteenth-century law texts as a term for renewing contracts, it was not a term for creation – it referred to newness. In the particularly entrenched religious atmosphere of sixteenth- and seventeenth-century Europe, doctrinal innovation was anathema. Some saw this kind of newness as affiliated with Puritanism or, worse, popery. Godin cites an extreme case from 1636 in which an English Puritan and former royal official, Henry Burton, began publishing pamphlets advocating against church officials as innovators. Burton used Proverbs 24:21 as his weapon: "My Sonne, feare thou the Lord, and the King, and *meddle not with them that are given to change*" (citation Godin's, emphasis added). The pot-stirring Puritan was accused of being the true "innovator," and sentenced to a life in prison and, in fact, a life without ears.

Innovation began to take root as a term associated with science and industry in the nineteenth century as the industrial revolution marched forward. Notably, the language of that period focused more on invention, particularly technical invention. According to Godin, several factors helped invention develop a prestigious and positive connotation, including the rise of a consumer culture, the increased numbers of patents, and the strong governmental focus on building labs for research and development.

"innovation," "data-driven," "AI," "growth"

When did the focus change from invention to innovation? Godin attributes this shift to a 1939 definition offered by the Austrian economist Joseph Schumpeter. Schumpeter defined invention as an act of intellectual creativity undertaken without any thought for its possible economic impact, while innovation happens when firms figure out how to craft inventions into constructive changes in their business models.

Over time, a new element was woven into the definition of innovation, which shifted the common understanding of the term to "bringing to market a new technology." In Godin's view, this was closely tied to government funding for research and development in laboratories and foundations. From the early 1950s until the 1980s, Godin says, innovation was understood as a process: theoretical research in labs provided an initial foundation; applications of that research were devised and developed; and those applications became commercialized products. Innovation was thought of as a packaged, predictable research product. According to Godin, government funding for these kinds of ventures directly corresponded to the emergence of this understanding of innovation.

How does today's fixation on innovation stack up against this history? From about 1870 to 1970, the US economy brimmed with newness. Since then, economist Tyler Cowen claims the forward march of technological progress has hit a dry spell, regardless of indications to the contrary:

> There's too much self-congratulations. . . . Americans have this self-image that we're the great inventors, [but] we've dropped the ball in many areas. We also see a lot of social tolerance – people confuse that with technological breakthroughs. [They have] a vague sense that things are getting better.[6]

According to Cowen, the internet counts as one of the biggest innovations in the past few decades. However, he believes that most of the economic gains from the internet will come from applications that have yet to be developed in areas like manufacturing. However, what has been achieved is a greater ability to manipulate information, which has an outsized effect on the lives and work of a relatively small segment of the population. These people happen to be those who spend the most time talking about innovation, such as journalists and academics.

Data-driven

"Data-driven" can be defined in many ways. For instance, it can be used as an adjective to refer to a process or activity that is spurred on by data, as opposed to being driven by mere intuition or personal experience (according to www.techopedia.com).

As Figure D.4 shows, there was relatively little development in the use of this term following the 2008 financial crisis. However, the interest tripled from 2013 to 2019.

In this connection, we can introduce another definition combined with a distinction between analogue and digital approaches to data. For many decades, data were treated in an analogue fashion, and this is still the case to some extent. For example, the use of Excel is still widespread. When we use the term "data-driven," we are in the digital world.

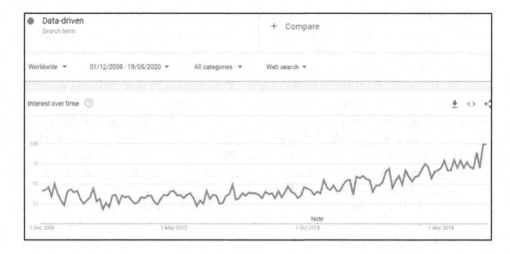

Figure D.4 Data-driven
Source: Google Trends (May 19, 2020).

Another key characteristic of "data-driven" is that it belongs to the big data mega-trend. As we saw in Chapter 3, a number of new companies have taken center stage as the largest companies in the world, including Google, Facebook, Amazon, Tencent, and Alibaba. Their key characteristic is a business model that heavily relies on data-rich digital platforms. These platforms are, above all, data-driven.

These very successful companies tend to further accelerate the use of data in their businesses. Once you have reached a critical mass of data, you are able to develop new services and new revenue streams over the top of the existing databases. This can be seen in Facebook's experience, as it initially obtained data through its platform for free. It retained ownership rights to that data, and it subsequently developed a number of services that provided a revenue stream.

A McKinsey report from 2016 states the following:

- "Data and analytics are changing the basis of competition. Leading companies are using their capabilities not only to improve their core operations but to launch entirely new business models. The network effects of digital platforms are creating a winner-take-most dynamic in some markets.
- Data is now a critical corporate asset. It comes from the web, billions of phones, sensors, payment systems, cameras, and a huge array of other sources – and its value is tied to its ultimate use. While data itself will become increasingly commoditized, value is likely to accrue to the owners of scarce data, to players that aggregate data in unique ways, and especially to providers of valuable analytics.
- Data and analytics underpin several disruptive models. Introducing new types of datasets ("orthogonal data") can disrupt industries, and massive data integration capabilities can break through organizational and technological silos, enabling new insights and models. Hyperscale digital platforms can match buyers and

"innovation," "data-driven," "AI," "growth"

sellers in real time, transforming inefficient markets. Granular data can be used to personalize products and services – and, most intriguingly, health care. New analytical techniques can fuel discovery and innovation. Above all, data and analytics can enable faster and more evidence-based decision making."[7]

A data-driven decision has significant value across all industries. However, one sector widely known to benefit from such insights is the airline industry. Southwest Airlines executives utilized targeted customer data to gain a deeper understanding of which new services would not only be most popular with customers but also most profitable. In so doing, the airline discovered that by observing and analyzing its customers' online behaviors and activities, it could provide different customer segments with the best rates for their needs in addition to an exemplary level of customer service. As a direct result of this emphasis on data-driven decisions, Southwest Airlines has seen its customer base and its brand loyalty grow steadily year after year.

AI

Colloquially, the term "artificial intelligence" (AI) is often used to describe machines (or computers) that mimic cognitive functions that humans associate with the human mind, such as "learning" and "problem solving." A more elaborate definition characterizes AI as "a system's ability to correctly interpret external data, to learn from such data, and to use those learnings to achieve specific goals and tasks through flexible adaptation."[8]

Frequently, when a technique reaches the mainstream, it is no longer considered artificial intelligence. This phenomenon is described as the AI effect.[9]

Tesler's theorem states that "AI is whatever hasn't been done yet."[10] Artificial intelligence can be classified into three types of systems: analytical, human-inspired, and humanized artificial intelligence. Analytical AI has characteristics consistent with cognitive intelligence. It generates a cognitive representation of the world and uses learning based on experience to inform future decisions. Human-inspired AI has elements from cognitive and emotional intelligence. It attempts to understand human emotions in addition to cognitive elements and considers them both in decision making. Humanized AI exhibits characteristics of all types of competencies (i.e., cognitive, emotional, and social intelligence), is able to be self-conscious, and is self-aware in interactions.

In the twenty-first century, AI techniques have experienced a resurgence following concurrent advances in computer power (Moore's law), the emergence of large amounts of data, and changes in our theoretical understanding. Physicist Stephen Hawking, Microsoft founder Bill Gates, and SpaceX founder Elon Musk have expressed concerns about the possibility that AI could evolve to the point where humans could not control it, with Hawking theorizing this could "spell the end of the human race." A group of prominent tech titans, including Peter Thiel of Amazon Web Services and Elon Musk have committed US$1 billion to OpenAI, a non-profit organization aimed at championing responsible AI development.

The sharp increase in searches for the combination of "AI" and "growth" tells the history of AI. The term was dormant for some years, but there has been a rather dramatic change in the last few years.

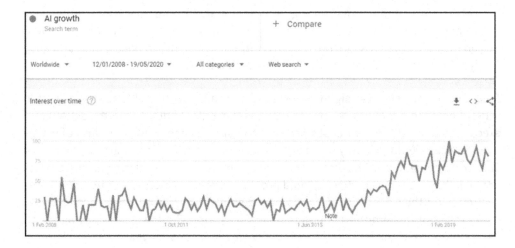

Figure D.5 AI growth

Source: Google Trends (May 19, 2020).

Growth

Growth strategies aim at winning a larger market share, even at the expense of short-term earnings. Four broad *growth strategies* are diversification, product development, market penetration, and market development. Common growth strategies in business include market penetration, market expansion, product expansion, diversification, and acquisition. The S-curve is a common way to illustrate growth.

In recent years, the strategic interest in growth has been tweaked somewhat, partly because of new, "strange" occurrences. One such occurrence was the advent of huge digital platforms, which meant a focus on questions like: What is your platform strategy? Are you multi-sided? Do you reap network benefits on both the supply and demand sides? Does your platform cater to innovative, over-the-top services? Some of the largest global companies are examples here, including Facebook, Google, Amazon, Alibaba, and Tencent.

A second type of strategic interest revolves around "digital rebels." At an early stage, these companies would have qualified, which is the same for many other companies. Also, consider Netflix versus Blockbuster.

A third example is companies that have changed their business models and in which the SaaS-model is gaining considerable traction. In short, this is a subscription-based model in which frontloaded customer payments are exchanged with recurrent subscriptions over a longer time span. Interestingly, many of the digital platforms referred to in this book pursue a combination of a diligent platform strategy, digital rebels, and the SaaS model.

Sean Ellis coined the term "growth hacker" in 2010.[11] In a blog post, he defined a growth hacker as "a person whose true north is growth. Everything they do is scrutinized by its potential impact on scalable growth." At the heart of growth hacking is a relentless focus on growth as the only metric that truly matters. Mark Zuckerberg had

"innovation," "data-driven," "AI," "growth"

this mindset while growing Facebook. While the exact methods vary from company to company and from one industry to the next, the common denominator is always to become measured by company growth (for example, revenue growth).

New customers typically hear about a product or service through their networks. After they use the product or service, they in turn share it with their connections. This loop of awareness, use, and sharing can result in exponential growth for the company.

The most relevant understanding of growth hacking frames the user-acquisition process using the "Pirate Funnel" metaphor. This embraces a six-stage funnel (awareness, acquisition, activation, retention, revenue, referral). It got its name from the abbreviation of the first letters of each word, which spell AAARR. Some consider growth hacking as simply an increase in the level of awareness by way of utilizing the social media, which is equal to the first "A" in the "AAARR"-model. However, true growth hacking implies success in all of the six phases of the "AAARR"-model.

An early example of "growth hacking" was Hotmail's inclusion of *"PS I Love You"* with a link for others to get the free online mail service. Another example was Dropbox's offer of more storage to users who referred the service to their friends.

Airbnb engaged in growth hacking through the coupling of technology with ingenuity. Airbnb realized that it could essentially hack the Craiglist.org scale and tap into its user base as well as its website by adding automated listing generators from Airbnb with the feature called "Post to Craigslist." The company's growth reflected a combination of clever thinking and technical know-how.

The combination of "data-driven" and "innovation"

Last, but not least, let us consider the combination of "data driven" and "innovation." The usage of this combination has increased exponentially in recent years – yet another reason for a book on data-driven innovation.

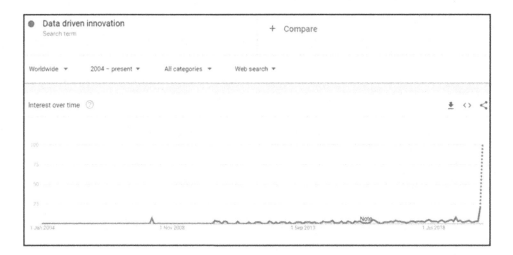

Figure D.6 Data-driven innovation

Source: Google Trends (May 19, 2020).

Notes

1 Cf. Geoffrey A. M. (2005). *Dealing With Darwin: How Great Companies Innovate at Every Phase of Their Evolution*. London: Penguin.
2 www.doblin.com.
3 Cf. Machiavelli, N. (2010). *The Prince*, edited by Michael Ashley. Campbell, CA: FastPencil, Inc., p. 21
4 Cf. https://mwi.usma.edu/1211-2/ (from 2016).
5 www.theatlantic.com/author/emma-green/.
6 Cf. http://businesshacker.co/what-is-innovation/.
7 McKinsey (2016). *The Age of Analytics: Competing in a Data-Driven World*, p. 8. https://www.mckinsey.com/business-functions/mckinsey-analytics/our-insights/the-age-of-analytics-competing-in-a-data-driven-world.
8 Cf. Kaplan, A. and Haenlein, M. (2019, January 1). Siri, Siri, in My Hand: Who's the Fairest in the Land? On the Interpretations, Illustrations, and Implications of Artificial Intelligence. *Business Horizons*, 62(1): 15–25.
9 Cf. McCorduck, P. (2004). *Machines Who Think* (2nd ed.). Natick, MA: A. K. Peters, Ltd.
10 Maloof, M. "Artificial Intelligence: An Introduction" (PDF). *georgetown.edu*, p. 37.
11 https://medium.com/growth-hacker/sean-ellis-on-growth-7d620cf4875f.

"innovation," "data-driven," "AI," "growth"

Bibliography

Andersen, M. M., Froholt, M. and Poulfelt, F. (2009). *Return on Strategy: How to Achieve It!* London and New York: Routledge.

Andersen, M. M. and Poulfelt, F. (2006). *Discount Business Strategy: How the New Market Leaders Are Redefining Business Strategy.* Hoboken, NJ: Wiley.

Andersen, M. M. and Poulfelt, F. (2014). *Beyond Strategy: The Impact of Next Generation Companies.* London and New York: Routledge.

Andersson, U., Dasi, A., Mudambi, R. and Pedersen, T. (2016). Technology, Innovation and Knowledge: The Importance of Ideas and International Connectivity. *Journal of World Business*, 51(1): 153–162.

Arthur D. Little and Match-Maker Ventures (2019). *The Age of Collaboration II: Start-ups + Corporates = Pain or Gain?* https://www.mckinsey.com/business-functions/mckinsey-analytics/our-insights/the-age-of-analytics-competing-in-a-data-driven-world.

Birkinshaw, J. M. and Ridderstrale, J. (2017). *Fast/Forward: Make Your Company Fit for the Future.* Stanford, CA: Stanford University Press.

Board of Innovation (2019). *Guide of Innovation Transformation.* https://www.boardofinnovation.com/guides/innovation-transformation-guide/.

Brynjolfsson, E. and McElheran, K. (2016). The Rapid Adoption of Data-Driven Decision-Making. *American Economic Review*, 106(5): 133–139.

CBInsights (2018). *State of Innovation: Survey of 677 Corporate Strategy Executives.*

Chesbrough, H. W. (2003). *Open Innovation: The New Imperative for Creating and Profiting from Technology.* Boston, MA: Harvard Business School Press.

Chesbrough, H. W. (2007). Why Companies Should Have Open Business Models. *MIT Sloan Management Review*, 48(2): 22–28.

Christensen, C. (1997). *The Innovator's Dilemma: When New Technologies Cause Great Firms to Fail*. Boston, MA: Harvard Business School Press.

Christensen, C. M. (2006). The Ongoing Process of Building a Theory of Disruption. *Journal of Product Innovation Management*, 23: 39–55.

Christensen, C. M., Raynor, M. and McDonald, R. (2015). What Is Disruptive Innovation? *Harvard Business Review*, 93: 44–53.

Curley, M. and Salmelin, B. (2018). *Open Innovation 2.0. The New Mode of Digital Innovation for Prosperity and Sustainability*. Cham: Springer.

Doz, Y. and Wilson, K. (2018). *Ringtone: Explaining the Rise and Fall of Nokia*. Oxford: Oxford University Press.

Dunning, D. (2011). The Dunning-Kruger-Effect: On Being Ignorant of One's Own Ignorance. *Advances in Experimental Social Psychology*, 44: 247–296.

Foss, N., Laursen, K. and Pedersen, T. (2011). Linking Customer Interaction and Innovation: The Mediating Role of New Organizational Practices. *Organization Science*, 22(4): 980–999.

Foss, N. and Pedersen, T. (2019). Microfoundations in International Management Research: The Case of Knowledge Sharing in Multinational Corporations. *Journal of International Business Studies*, 50: 1594–1621.

Foss, N., Pedersen, T., Pyndt, J. and Schultz, M. (2012). *Innovating Organization & Management: New Sources of Competitive Advantage*. Cambridge: Cambridge University Press.

Geoffrey A. M. (2005). *Dealing With Darwin: How Great Companies Innovate at Every Phase of Their Evolution*. London: Penguin.

Gifford, J. (2012). *Blindsided. How Business and Society Are Shaped by Our Irrational and Unpredictable Behavior*. London: Marshall Cavendish Business.

Gruber, M. and Tal, S. (2017). *Where to Play*. London: Pearson.

Hoffman, R. and Yeh, C. (2018). *Blitzscaling: The Lightning-Fast Path to Building Massively Valuable Companies*. New York: Currency.

Innosight (2018). *2018 Corporate Longevity Forecast: Creative Destruction is Accelerating*.

Kahneman, D. (2011). *Thinking Fast and Slow*. New York: Allan Lane.

Kaplan, A. and Haenlein, M. (2019, January 1). Siri, Siri, in My Hand: Who's the Fairest in the Land? On the Interpretations, Illustrations, and Implications of Artificial Intelligence. *Business Horizons*, 62(1): 15–25.

Lashinsky, A. (2012). *Inside Apple. The Secrets Behind the Past and Future Success of Steve Jobs's Iconic Brand*. London: John Murray.

Laursen, K. and Salter, A. (2006). Open for Innovation: The Role of Openness in Explaining Innovative Performance Among UK Manufacturing Firms. *Strategic Management Journal*, 27(2): 131–150.

Maloof, M. "Artificial Intelligence: An Introduction" (PDF). *georgetown.edu*, p. 37.

Markides, C. C. (2008). *Game-Changing Strategies: How to Create New Market Space in Established Industries by Breaking the Rules*. San Francisco, CA: Jossey-Bass.

McCorduck, P. (2004). *Machines Who Think* (2nd ed.). Natick, MA: A. K. Peters, Ltd.

McKinsey (2016). *The Age of Analytics: Competing in a Data-Driven World*. https://www.mckinsey.com/business-functions/mckinsey-analytics/our-insights/the-age-of-analytics-competing-in-a-data-driven-world.

OECD (2015). *Data-Driven Innovation: Big Data for Growth and Well-Being*. OECD: Paris.

Osterwalder, A. and Pigneur, Y. (2010). *Business Model Generation: A Handbook for Visionaries*. Hoboken, NJ: Wiley.

Osterwalder, A., Pigneur, Y., Smith, A. and Etiemble, F. (2020). *The Invincible Company: How to Constantly Reinvent Your Organization with Inspiration From the World's Best Business Models (Strategyzer)*. Hoboken, NJ: Wiley.

Rosenzweig, P. (2007). *The Halo Effect . . . and the Eight Other Business Delusions that Deceive Managers*. New York: Free Press.

Siilasmaa, R. (2019). *Transforming Nokia: The Power of Paranoid Optimism to Lead Through Colossal Change*. New York: McGraw-Hill Education.

Syed, M. (2019). *Rebel Ideas: The Power of Diverse Thinking*. London: John Murray.

Tvede, L. (2019). *Supertrends*. Copenhagen: Politikens Forlag.

Von Hippel, E. (2016). *Sources of Innovation*. New York: Oxford University Press.

Index

Note: Page numbers in *italics* indicate a figure and page numbers in **bold** indicate a table on the corresponding page.

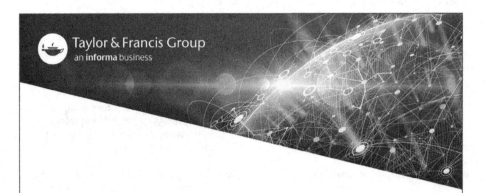

Printed in the United States
By Bookmasters